EPISODES FROM THE HISTORY OF
THE RARE EARTH ELEMENTS

Chemists and Chemistry

VOLUME 15

A series of books devoted to the examination of the history and development of chemistry from its early emergence as a separate discipline to the present day. The series will describe the personalities, processes, theoretical and technical advances which have shaped our current understanding of chemical science.

The titles published in this series are listed at the end of this volume.

EPISODES FROM THE HISTORY OF THE RARE EARTH ELEMENTS

Edited by

C. H. Evans

University of Pittsburgh,
Pittsburgh, Pennsylvania, U.S.A.

Kluwer Academic Publishers

Dordrecht / Boston / London

A C.I.P. Catalogue record for this book is available from the Library of Congress.

ISBN 0-7923-4101-5

Published by Kluwer Academic Publishers,
P.O. Box 17, 3300 AA Dordrecht, The Netherlands.

Kluwer Academic Publishers incorporates
the publishing programmes of
D. Reidel, Martinus Nijhoff, Dr W. Junk and MTP Press.

Sold and distributed in the U.S.A. and Canada
by Kluwer Academic Publishers,
101 Philip Drive, Norwell, MA 02061, U.S.A.

In all other countries, sold and distributed
by Kluwer Academic Publishers Group,
P.O. Box 322, 3300 AH Dordrecht, The Netherlands.

Printed on acid-free paper

Printed in the Netherlands

TABLE OF CONTENTS

Preface

At the beginning of the "Biochemistry of the Lanthanides" (Plenum, NY, 1990), I made some remarks concerning the history of the rare earth elements and commented that there was "enough material here for an historian of science to write an instructive book on the identification of the lanthanides in its own right". Little did I think when writing these words that just a few years later I would be editing such a volume which, in recognition of the growing industrial, biological and medical importance of these elements, has been extended to include a second section, devoted to their application.

Interest in the history of these elements has been increasing in recent years, a trend accelerated by the bicentenary, in 1987, of the discovery of the first rare earth mineral. Although there exist scattered articles and book chapters pertaining to the history of the rare earths, there is only one other serious volume entirely devoted to historical matters concerning these elements: "Two hundred year impact of rare earths on science" edited by K.A. Gschneidner and L.R. Eyring (Elsevier, Amsterdam, 1988). This book, most chapters of which are highly technical and scientific in nature, contrasts with the present volume which concentrates more on matters of historiography and contextual interpretation. Taken together, they constitute comprehensive package both from the point of view of history and science.

All agree that the discovery of the rare earth elements began in 1787 when Carl Axel Arrhenius, a Swedish army officer, happened upon a new mineral which he named ytterbite after the nearby village of Ytterby. From this mineral the Finnish chemist Johan Gadolin isolated ytterbia (Y_2O_3) in 1794, thus fuelling an era of discovery that engaged many of the greatest chemists of the age. The Scandinavian origins of the rare earths are nicely reflected in the first three chapters of this volume, whose authors come from Finland and Sweden. Their contributions provide original insights into the work of Gadolin, Hisinger, Berzelius and Mosander, among others, which led to the identification of the first five rare earth elements.

The subsequent chapter summarizes the events of approximately 1850-1900 during which time technologic al and conceptual advances facilitated the discovery of all but two of the remaining rare earth elements. However, the chain of events set in motion by the discovery and chemical analysis of ytterbite, later re-named gadolinite, did not end until Marinsky unequivocally identified the final rare earth element promethium (Pm) in 1947. It is a particular privilege to have in this book a valuable chapter by Professor Marinsky telling in his own words how this discovery came about.

Part I of this book provides much novel scholarship on the discovery of the rare earth elements. It is clear from these chapters that the isolation of these elements not only has a fascinating internal history, but also contains many episodes of broad significance to the wider history of chemistry and, indeed, to the history of science as a whole. The importance of technical innovation and theoretical advance to the process of chemical discovery are, for example, resoundingly exemplified by rare earth history. Light is shed upon the factors which determined the chemist's choice of problems and the robust empiricism used in many cases to solve them. And, as exemplified in Dr. Helge Kragh's masterful chapter, these elements provide wonderful case studies for the analysis of priority disputes in scientific discovery.

Rare earth elements provided important challenges for Mendeleev's periodic table, which he revised on many occasions to accommodate these troublesome individuals. Yet this interplay proved profitable and helped to validate Mendeleev's system, which predicted the existence of "*eka-boron*" with properties corresponding to those of scandium. Moreover, the discovery of Pm heralded the first occasion on which an element which did not exist in nature was predicted, generated and discovered.

Part II of this book is devoted to the history of applying rare earths for practical purposes. Industrial application of the rare earths is inextricably linked to the Austrian chemist Carl Auer von Welsbach. It is difficult to overstate the importance of this individual to the commercial application of the rare earths, yet there is little biographical information to be found in the English language. The chapter by Dr. Edwin Baumgartner, who worked until recently at the Treibacher Chemical Works founded by Auer, is thus of particular value. According to this chapter, the origin of the rare earth industry can be traced to November 4, 1891, when Auer's gas mantle was used to illuminate a Viennese café. Baumgartner's analysis of Auer's life and work illustrates nicely how the search for practical utility, the demands of commerce and the process of scientific discovery can sometimes synergise.

China has the world's largest reserves of rare earth ores and is at the forefront of attempts to diversify and extend the industrial use of these elements. These efforts are spearheaded by scientists at the Baotou Research Institute in Inner Mongolia. A detailed chapter by two members of this institute describes the short history of this work and the present huge efforts to realise the full commercial potential of these elements.

Not all applications of the rare earths have economic ends. Improvements in methods for determining rare earth elements in geological samples, coupled to the advent of plate tectonic theory, has led to an explosion in the use of these elements in geochemistry. As comprehensively described by Drs. Edward Lidiak and Wayne Jolly, this continues to be a fruitful and ever growing area of application for the rare earth elements.

The last two chapters are biological in nature. Since the 1960s, it has become increasingly appreciated that trivalent lanthanide ions make excellent isomorphic replacements for calcium ions in biology and biochemistry. Being more informative than calcium, the lanthanides have been widely used to probe calcium sites in biology. A major application has been to study calcium binding sites on the surfaces of cells, particularly those which undergo stimulus-coupled responses. One of the pioneers of this field, Dr. Arthur Weiss, describes how these developments came about. Finally, there is a chapter analysing some of the various medical uses to which rare earth elements have been put.

In editing these chapters I have been at pains to make as few changes as was reasonably possible to the authors' original texts. This has inevitably led to a variety of styles of exposition, ranging from focussed, personal, first hand accounts to detached broad analyses. Nevertheless I believe that a book such as this, covering a wide range of topics dating from over 200 years ago to the near present, gains from this practice. Allowing individual authors free reign has also led to inevitable overlap between chapters but, again, I feel this is to the good. It permits each chapter to read separately and to be understood without reference to other chapters, while at the same time reinforcing the essential links that exist between the various facets of rare earth history.

By forming part of Kluwer's series of historical studies on "Chemistry and Chemists", I hope that this volume will make its information accessible to the wider audience and lead to greater interest in the rare earth elements and their history.

Acknowledgements

The origins of this book lie with an invitation from Drs. Janjaap Blom who was, at that time, the Acquisition Editor for Kluwer Academic Press. He had read a short article entitled "The Discovery of the Rare Earth Elements", which I had written for *Chemistry in Britain*, and suggested that I might be interested in editing a book on the subject. With the completion of this volume, I am eager to express my gratitude to Drs. Blom for suggesting this project and encouraging me to undertake it. (I must admit there were times during the project's gestation when I wasn't quite so grateful).

There are relatively few scholars working in the suject area covered by this book. It was therefore important that those individuals whom I invited to contribute chapters would actually agree to do so. I would like to thank all the authors for their contributions and particularly for providing so much previously unpublished new scholarship as opposed to rehashes of existing information. Without their carefully written chapters this book would not, of course, exist. In this context, I thank Professor R.J.P. Williams of Oxford University for suggesting that I include a chapter on geochemistry, and Dr. Barry Kilbourne of Molycorp Inc. for putting me in touch with Dr. Edwin Baumgartner, then of Treibacher Schleifmittel, concerning the biography of Auer von Welsbach.

As usual, I am indebted to family, friends and colleagues for putting up with me while I engaged in yet another time-consuming distraction. Much of the chaos I generate in the pursuit of such enterprizes ends up on the desk of my secretary, Mrs. Lou Duerring. It goes almost without saying that without her intervention this book would still be unfinished. I am particularly grateful to Lou for rising to the occasion when,at the eleventh hour, it was decided to revise the book into a camera-ready format in order to speed up publication. Thanks Lou, for becoming an instant expert in desk-top publishing!

Introduction

I am willing to venture the opinion that the history of the rare earths is more fascinating, and illuminates more areas of chemical progress, than the history of any other group of elements. Such a position resonates with the assessment of the mineralogist Flint who is reported to have claimed that ytterbite was maybe the most significant single mineral in the history of inorganic chemistry. Perusal of the first part of this book will provide much support for such statements.

Asccording to IUPAC nomenclature, the rare earth series comprises elements 21 (Sc), 39 (Y) and 57 (La) to 71 (Lu). The first of these to be discovered was yttrium. It began in the summer of 1787 when the Swedish army office, Carl Arrhenius found a heavy, black mineral which he named ytterbite after the adjacent village of ytterby. In 1794 the Finnish chemist Gadolin isolated from ytterbite a new earth, which we now know to be yttria - Y_2O_3. Although yttrium was not obtained in pure, elemental form, Gadolin's 1794 isolation is taken to represent the discovery of the first rare earth. Production of the first of this series in elemental form had to await Mosander's reduction of ceria to cerium in 1826.

Gadolin's isolation of yttria represented the first of a number of rare earths that were extracted from ytterbite, later renamed gadolinite by Ekeberg. As indicated in figure 1, and as explained in detail in the first part of this book, chemical analysis of gadolinite was responsible for the discovery of 9 members of the rare earth series. These are sometimes referred to as the "yttric" group. Discovery of all but one of the remaining 8 elements can be traced directly or indirectly to the mineral cerite. These elements are sometimes referred to as the "ceric" group.

Cerite, originally called tungsten, served as the starting material for the isolation in 1803 of ceria (Ce_2O_3) by Berzelius and Hisinger in Sweden and, independently, by Klaproth in France. In chapter 2, Professor Jan Trofast provides details of the ensuing priority dispute - the first of several along the pathway of rare earth discovery.

As shown in figure 1, cerite brought forth not only cerium, but also lanthanum and didymium; the latter was shown to be a mixture of praseodymium and neodymium. Progress in this direction was aided by the discovery of more abundant rare earth minerals, such as samarskite. Although identified in the USA in 1878, it was named after the Russian mining engineer Samarsky who had earlier found this mineral in the Urals. Rare earth chemists were also aided by Auer von Welsbach's importation of large amounts of monazite sand as a commercial source of rare earths (Chapter 7).

In reading accounts of the discoveries of the rare earth elements, one is struck by the ways in which technology and, later, theory both aided and, in some cases, hindered progress. Up until the advent of ion exchange chromatography in the 20th century, chemists were forced to use inefficient and laborious techniques such as fractional precipitation and crystallisation, to separate rare earths. Because the chemical properties of the various members of this series are so similar, their separation and purification was an enormous undertaking which would have difficulty attracting grant support today. Tens of thousands of repetitive fractionations were required to produce tiny amounts of purified material.

Problems of purification were compounded by the lack of good methods for identifying elements and assessing their purity, as well as complete ignorance of the number of different rare earths that actually existed. Spectral analysis, introduced by Bunsen and Kirchhoff in 1859, and the development and refinement of Mendeleev's periodic system provided partial answers to these problems. Their impact is clear. In the 60 years or so prior to the introduction of these two advances, only 6 rare earths had been identified and one of these, didymium, did not exist. During the subsequent 50 years, the list of rare earth elements jumped to 15, only lutetium and promethium being absent.

The interplay between Mendeleev's periodic system and the rare earth elements is an interesting one, worthy of deeper analysis by future scholars. For instance, Mendeleev's system predicted the existence of scandium and accurately foretold several of its properties. In this sense, the rare earths facilitated acceptance of Mendeleev's proposed periodic system. Nevertheless, the other rare earth elements severely challenged his system, because he didn't know where to place them. Many modifications were made simply to accommodate these elements. Such difficulties continues to dog Mendeleev until his death. Meanwhile, spectroscopic analysis was proving to be as much a hindrance as a help. Because of the spectral complexity of the rare earths, the difficulty of preparing them to high purity and the unreliability of spectroscopy in inexperienced hands, rare earth elements were being discovered in ridiculous numbers; over 70 in just a few years. Clearly something was wrong, but without knowing how many rare earth elements actually existed, it was hard to know how to resolve the issue.

Relief came in the form of Moseley's finding, in 1913, that the atomic number of an element could be determined from its X-ray spectrum. Calculations based upon these premises quickly revealed that all possible rare earth elements had already been discovered, except one - element 61. Here is a good example of theory coming to the aid of practice. Thanks to Moseley's theory, effort need not be wasted looking for non-existent elements; instead, resources could be re-directed towards the discovery of the final missing member. This was finally achieved in 1947 (Table 1).

Of interest to historians of science is the resistance to Moseley's ideas from certain chemists. As Dr. Kragh points out in chapter 5, this is an illustrative case of a "boundry conflict" between disciplines, complicated by the clash of cultures occurring between the older, practical, experimentalists and the new theoreticians. This chapter also provides an illuminating analysis of the priority dispute between Auer von Welsbach and Urbain over the discovery of element 71. Also pointed out is the possibility that both of these chemical giants may have been beaten to the punch by Charles James at the University of New Hampshire. For some reason, perhaps timidity in the face of such a viscious priority battle by two such noted individuals, James did not pursue his claim.

Discovery of the final rare earth was impeded by the fact that it is an unstable, radioactive element that exists naturally on earth in only the minutest of quantities. There were several false starts in the search for element 61, including some strange behaviour on the part of two Florentine scientists who claimed to have discovered the element in 1924 but, for reasons that are completely obscure, to have locked the manuscript away in the Academia dei Lincei. This odd circumstance was only revealed in 1926 when scientists at the University of Illinois claimed to have isolated the same element. In the event, all such claims were groundless. Element 61 was unequivocally isolated for the first time by Marinsky's group in 1947; Professor Marinsky tells how this came about in chapter 6. Particularly impressive is the manner in which the senior scientist associated with the project, Professor Coryell, removed his own name from the paper announcing the discovery, so as not to detract from the accomplishments of the other, more junior investigators.

The first practical application of a rare earth seems to have been in the unlikely role as a medicine. An 1854 paper by Professor Simpson of Edinburgh, Scotland, extolled the benefits of using cerium nitrate to treat the reflex vomiting of pregnancy. He later switched to recommending cerium oxalate which, in a short space of time, became highly prescribed for the treatment of all types of vomiting, including sea-sickness, as well as for coughs and even various nervous disorders such as epilepsy and chorea. The popularity of cerium oxalate waned almost as quickly as it had arisen, and it is no longer used for this purpose Since that time, however, various rare earth salts have been used as antimicrobial agents for the treatment of infectious diseases, as anticoagulants, and as antiinflammatory agents. Today, cerium nitrate is used to treat burns and GdDTPA is used as a contrast enhancing agent in magnetic resonance imaging.

By a route that is completely unrelated to medical use, rare earths have become widely used as experimental tools in cell biology and biochemistry. Trivalent cations of the rare earth metals serve as informative, isomorphous replacements for calcium ions which, although ubiquitous and extremely important in biology, are spectroscopically dead. One aspect of this application, as a way to probe calcium-binding to the surface of muscle cells, is described in chapter 10 by one of the pioneers of this field.

Another important scientific application of the rare earth elements is in the area of geochemistry (chapter 9). Here again we see how such a use requires the combination of an underlying theory, in this case plate tectonics, with the appropriate technological sophistication, in this case the ability to measure tiny concentrations of rare earth elements to high accuracy.

Industrial application of the rare earths began on November 4, 1891 outside the Opern café in Vienna (Chapter 7). Carl Auer von Welsbach had invented an incandescent gas mantle comprising a thin fabric inpregnated with thorium oxide and cerium oxide. After the demonstration in Vienna, this form of illumination rapidly gained widespread use and the city of Bombay used it for public lighting. As well as initiating the industrial use of the rare earths, Auer von Welsbach isolated praseodymium and neodymium from didymium and discovered lutetium, which he called cassiopeium . The priority dispute between Urbain and Auer von Welsbach concerning element 71 had already been mentioned.

Auer von Welsbach also invented a lighter flint based upon alloy of iron and cerium, which became a huge commercial success. Because of these successes, the chemical works founded by Auer in Treibach, Austria is still in operation. Auer required an abundant source of rare earth minerals, which he satisfied by realising the properties of an ore called monazite which occurred in the New World and was, until that time, used as ballast by sailing ships.

China has the world's largest reserves of rare earth ores and is understandably eager to exploit them commercially. As described in chapter 8, the history of rare earth applications in China is relatively short. Their production in China did not begin until the 1950s and most of the major deposits were not identified until the 1960s. Research programmes to improve extraction procedures and processing have been in place since this time. New uses are also being actively sought not only in traditional areas of technology, but also in such areas as agriculture, where the Chinese claim that the use of fertilizers containing the rare earths improves growth and increases yields.

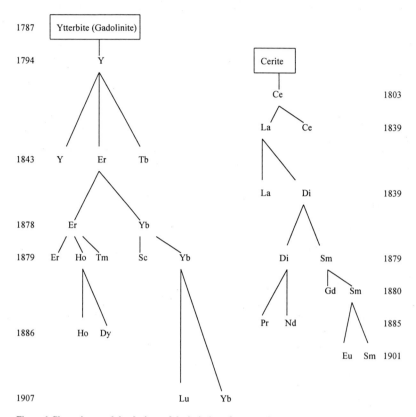

Figure 1 Chronology and dendrology of the isolation of rare earths.

As this diagram illustrates, the discovery of each of the stable rare earth elements can be traced to a "yttric" or "ceric" origin. Progress was delayed by the rarity of ytterbite (gadolinite) and cerite, a deficiency that was remedied by the discovery of Samarskite and, later, Monazite which are abundant rare earth ores.

The rare earths were not initially isolated as pure metals but, rather, as impure preparations of their salts. Mosander purified the first rare earth compound, lanthana (La_2O_3) and later prepared a rare earth (Ce) in metallic form for the first time.

With the exception of didymium, when a previously pure rare earth was shown to contain an additional earth, or earths, one of the new elements retained the original name of the mother element, while the other daughter element was provided with a new name.

Table 1 Discovery and etymology of the rare earths

Rare Earth	Year of Discovery	Discoverer	Origin of Name	Reference
Yttrium (Y)	1794	Gadolin	Ytterby, a village near Stockholm	Chapter 1
Cerium (Ce)	1803	(i) Berzelius & Hisinger (ii) Klaproth	Ceres, an asteroid discovered in 1801	Chapter 2
Lanthanum (La)	1839	Mosander	To lie hidden (Greek)	Chapter 3
Erbium (Er)	1843	Mosander	Ytterby	Chapters 3 & 4
Terbium (Tb)	1843	Mosander	Ytterby	Chapters 3 & 4
Ytterbium (Yb)	1878	Marignac	Ytterby	Chapters 4 & 5
Holmium (Ho)	1879	Cleve	Stockholm (Latin)	Chapter 4
Thulium (Tm)	1879	Cleve	Northernmost region of the inhabitable world (Latin)	Chapter 4
Scandium (Sc)	1879	Nilson	Scandanavia (Latin)	Chapter 4
Samarium (Sm)	1879	Lecoq DeBoisbaudran	Samarskite, an ore named after the Russian army officer Samarsky	Chapter 4
Gadolinium (Gd)	1880	Marignac	Johan Gadolin	Chapter 4
Praseodymium (Pr)	1885	Auer	Green twin (Greek)	Chapters 4 & 7
Neodymium (Nd)	1885	Auer	New twin (Greek)	Chapters 4 & 7
Dysprosium (Dy)	1886	Lecoq DeBoisbaudran	Hard to get at (Greek)	Chapter 4
Europium (Eu)	1901	Demarçay	Europe	Chapter 4
Lutetium (Lu)	1907	(i) Auer (ii) Urbain	Paris (Latin)	Chapters 5 & 7
Promethium (Pm)	1947	Marinsky Glendenin Coryell	Prometheus, Greek God	Chapter 6

ESPISODES FROM THE HISTORY OF RARE EARTH ELEMENTS

PART I - DISCOVERY

Lieutenant Carl Axel Arrhenius. Discoverer in 1787 of ytterbite, later renamed gadolinite, the mineral which launched the quest for the rare earth elements.

CHAPTER 1

WHAT DID JOHAN GADOLIN ACTUALLY DO?

PEKKA PYYKKÖ
Department of Chemistry, University of Helsinki
Et. Hesperiankatu 4, SF-00100 Helsinki, Finland

OLLI ORAMA
Department of Chemistry, University of Helsinki,
Vuorikatu 20, SF-00100 Helsinki, Finland

1.1. Johan Gadolin: a short biography and bibliography

Johan Gadolin (5 June 1760 - 15 August 1852) was born in Turku (Åbo) in Finland, which then was an integral part of Sweden. He first studied at the Royal Academy at Åbo in 1775 and later at Uppsala (1779-83), receiving a degree there in 1781. His thesis was called "*De analysi ferri*" and thesis advisor, Torbern Bergman (1738-84). He held professorial positions at Åbo from 1785 until his retirement in 1822.
Gadolin's papers are listed in the reprint and memorial volume by Hjelt and Tigerstedt (1910). Further details can be found in Asimov (1976), Enkvist (1972), Klinge et al. (1987), Mäkitie (1982), Pyykkö and Orama (1988), Sjöblom (1960a,b), Tigerstedt (1899) and Toivanen (1980).

1.2. The discovery of the mineral

The Ytterby lanthanide mineral was discovered by Lieutenant Carl Axel Arrhenius and announced by Geijer (1788). The paper is short enough to be entirely reproduced on the cover of the "Abstracts and Programme" of the 2nd International Conference on Lanthanides and Actinides in Lisbon, Portugal in 1987. Geijer reports physical properties (black, asphalt- or coal-like colour, high density of 4.223, lack of magnetism of the stone or of the oxide, magnetism of the reduced metal). Preliminary experiments with a blowpipe and with calcination are mentioned. The only preliminary conclusion is a proposal that the sample might contain tungsten. A later paper on "Pitchblende" in Rinman's "Bergverks Lexicon" is quoted by Gadolin (1794, 1796).

1

C. H. Evans (ed.), Episodes from the History of the Rare Earth Elements, 1–12.
© 1996 Kluwer Academic Publishers. Printed in the Netherlands.

1.3. Gadolin's analysis: The 1794 and 1796 papers

A rather more detailed analysis was reported by Gadolin (1794, 1796) in the nearly identical Swedish and German versions of the same paper, after he had obtained a small sample from "Captain Arrhenius". The organization and contents of the paper are as follows:

§1.

Physical appearance: This mineral occurred in a red feldspar as rounded aggregates or as parallel sheets. The density (4.028 compared to water, see Gadolin (1792)), the grey-green colour of the ground powder and an idea of the hardness are given.

§2.

Gadolin managed to melt, with difficulty, the stone in a blow-pipe into a black, bubbly slag, in contrast to Geijer, who failed to achieve this, and Rinman, who obtained a yellow glass. Melts were also obtained with sodium hydroxide [Soda-Alkali] borax and potassium hydrogen tartrate ["*Sal microcosmicus*" (Swe.)/ "*Weinsteinsalz*" *(Ger.)]*. A mixture of the ground stone with two parts of saltpetre did not explode in a glowing crucible. Leaching with water and drying yielded a brown powder of the original weight. The metallic residue of Geijer and Arrhenius is quoted.

§3.

A) A ground sample of the stone was dissolved in concentrated nitric acid, giving a white, insoluble powder in a greenish solution. Evaporation of water gave a jelly, redissolved by water.

B) The brown powder, obtained by melting in saltpetre, behaved similarly.

C) Solutions in hydrochloric acid also gave a greenish solution and an insoluble powder. Some heat and a little gas were generated. A smell, not unlike the "hepatic", was detected. Strong acid gave a thick, oily solution. The final colour was brown-yellow.

D) Hydrochloric acid also attacked larger pieces of the stone, giving a white remnant stone and a similar solution.

§4.

The insoluble residue was dried and had a weight of about 1/3rd of the original. It could not be melted in a blow-pipe, not even with feldspar. With sodium carbonate a clear glass was obtained, with evolution of gas. On cooling, the glass became milky. Saturation with saltpetre gave a white pearl of enamel. Borax gave a clear glass. *Sal microcosmicus* gave no reaction.

It consequently follows that the part, insoluble in acids, mainly contains silica, (underlined by translators), but that it also contains something else less capable of reacting with alkali carbonate. Undoubtedly, some feldspar was present. Whether a metal oxide, insoluble in acids, could also be present was investigated as follows: 1 part of the insoluble residue was mixed with 3 parts of crystalline sodium carbonate and calcined in a crucible for one hour. A spume, that was mostly soluble in water, resulted. The insoluble part, rinsed and dried, had 1/5th of the weight of the entire residue. From it, sulphuric acid could extract a slight amount of aluminium oxide.

The remnant was mixed with 3 parts of sodium carbonate, calcined and dissolved in water. The residue that remained undissolved had about 1/20th of the mass of the original residue, but it behaved similar to the way it had before alkali melting; i.e. with sodium carbonate it melted, with gas evolution, into a pearl which was clear when warm but opaque when cool. Once, a small silvery metal fragment was observed in the pearl. A slight addition of copper oxide gave the glass a ruby-red colour. This suggests that the residue may have contained some tin oxide.No tin layer could be obtained on an iron wire inserted into the glass, however.

After calcination with sodium carbonate the water-soluble part formed a jelly with hydrochloric acid, behaving just like pure silica [water glass].

§5.

A) The solution in hydrochloric or nitric acid (§3), containing 2/3rd of the weight of the earth, gave in pure potash solution first a brownish, then a white precipitate which, rinsed and dried in warm air, weighed somewhat more than the entire earth before dissolution.

B) With ammonia, the same solution gave a grey precipitate, turning dark brown when heated. After addition of a slight excess of ammonia, all earth was separated from solution, and no further precipitation resulted with alkali [*feuerfeste Laugensalz*]. Hence these solutions contained no observable traces of tungsten oxide [*Schwererde*] nor of calcium or magnesium oxide; this conclusion will shortly be confirmed. The precipitate with ammonia had, after rinsing and drying, somewhat more than the mass of the earth. Its mass was reduced through calcination to 2/3rd of the original; so was that of the precipitate (A).

Both precipitates behaved similarly in blowpipes. They darkened during heating, hardened at higher temperatures and became slaggy on the surface, but did not melt. With borax they yielded a clear, greenish or black glass, depending on the proportions, and with *Sal microcosmicus* a clear glass, containing white particles above the phase separation. These did not dissolve by melting with sodium hydroxide [*Alkali Sodae*]. The white precipitate, finally obtained by potash solution, collected and dried became bleak red-grey as it dried. By blow-pipe in the presence of borax it gave a clearer glass that in the outer flame obtained a hyacinth colour; it finally became opaque when blown for a long time at the extremity of the flame. It became clear again in the blue flame.

C) Macquer's "*Blutlauge*" ($K_4[Fe(CN)_6]$) separated from the mentioned solutions a large, blue precipitate in which a slight amount of white powder could be seen. Thus 100 parts of the powdered earth yielded 41 parts of precipitate. This corresponds to 12 parts iron oxide if a pure calcined iron oxide is dissolved in the same acid and precipitated by the same "*Blutlauge*".

With potassium hydroxide the clear solution that was not precipitated by $K_4[Fe(CN)_6]$, gave a white precipitate that, after rinsing and drying in warm air, weighed 87 parts. This was partially dissolved by concentrated potassium hydroxide [caustic vegetable alkali]. Addition of vitriolic acid to the obtained alkaline solution gave a white precipitate, with a cubic crystal habit, weighing 29 parts. It was soluble in vitriolic acid and gave, by evaporation, crystals of alum. The part not dissolved by alkaline potash solution was an unknown earth that shall be described more closely below. After rinsing and drying, its mass was 38 parts.

Using the weights after heat treatment, the various constituents of 100 parts of the black earth contain approximately:

31 parts silica
19 parts aluminium oxide
12 parts iron oxide
<u>38</u> parts of an unknown earth.
100

Sometimes the undissolved silica appeared to contain some tin oxide, which probably formed the white particles in the $K_4[Fe(CN)_6]$ precipitate (see next section). The blowpipe behaviour also suggested some manganese dioxide.

§6.

The brown, dried precipitate of §5A was analysed further as follows:

A) It could be slowly dissolved in distilled, weak acetic acid, requiring 120 parts thereof. Some brown-black powder remained; it consisted of iron oxide, with some aluminium oxide.

B) Dilute sulphuric acid dissolved it entirely, developing initially both gas and heat. The brown powder in the bottom disappeared by gentle heating. The solution was pale green. A piece of blank iron, immersed in this solution, was attacked by the remaining acid. No metallic precipitate [consisting of nobler metals] appeared on it. An immersed copper sheet did not change its colour either. If both iron and copper were immersed, a thin white metal layer appeared on the latter.

C) Nitric acid similarly dissolved it, with the evolution of gas, into a clear solution which, by evaporation, became viscous and later formed a jelly. By further heating it foamed, emitting red-brown fumes, and finally became a dark-brown residue containing white particles. The same procedure was repeated with the same amount of nitric acid (2 parts of acid to one part of precipitate). Finally, 2 parts nitric acid and 4 parts water were added to the dark-brown powder. The solution was boiled and diluted with 10 parts water. Upon filtration the solution left a rust-brown powder, 4 per cent by weight of the original precipitate. The filtered solution gave with potash solution first a white precipitate which dissolved again as the acid approached neutrality. When further alkali was added, the solution became yellow, turning later dark-brown, opaque, and precipitated a dark-brown powder, increasing upon boiling. Separated and dried, these black, cubic crystals weighed 16 per cent of the original material.

The filtrate was clear, despite containing dissolved iron oxide. With potash solution it gave a white, slightly reddish precipitate, weighing 80 per cent of the original material when rinsed and dried. It was dissolved in nitric acid, precipitated again with potassium hydroxide and rinsed with water. While it was still wet, concentrated potassium hydroxide was added. Upon heating it leached 20 per cent of its substance which behaved as pure aluminium oxide. (This percentage is slightly below that in the previous paragraph, due to previous coprecipitation with iron oxide).

The remaining part of the powder that was not dissolved by the potassium hydroxide had properties similar to the unknown earth of §5.C, except that it still contained some iron oxide, making it slightly reddish. It was purified by dissolution in acid and precipitation by $K_4[Fe(CN)_6]$.

§7.

The unknown earth had the following properties:

Even at the highest temperatures obtained with the blow-pipe, it remained white and did not melt [the melting point of Y_2O_3 is 2410°C].

Borax dissolved it into a clear, colourless glass.

With sodium ammonium hydrogen phosphate it yielded a glass, which remained colourless in the blue flame of the blowpipe but became milky in the outer flame.

With boric acid it gave a white enamel.

It did not dissolve into sodium hydroxide upon melting.

High-temperature treatment in a crucible placed within a coal-oven did not give any change under conditions which reduced iron. Melting with borax under the same conditions also failed to yield any metal.

In wet chemistry this earth easily dissolved in most acids giving clear, colourless solutions.

Its *sulphuric acid* solution did not yield bigger crystals. During evaporation a white powder was obtained. Its taste, after rinsing, was sweet-sour. If the solution was fully dried, a white powder was obtained that, in its appearance and slight solubility, rather resembled selenite but, in its taste, alum. An excess of sulphuric acid readily dissolved it in water.

With *nitric acid* a clear solution was obtained. Upon evaporation it formed a jelly and could not be brought to crystallization.

A solution in *hydrochloric acid* showed no tendency to crystallize either, instead becoming viscous. From all these solutions the earth could be precipitated by ammonia. The precipitate was insoluble in alkaline solutions.

Carbon dioxide reacted with this earth in the sense that the precipitate obtained with alkali carbonates evolved gas when treated by acids. This precipitate was only slightly soluble in carbonated water (1 part in 7000).

Phosphoric acid, obtained from phosphorus by slowly burning it in air, did not precipitate the earth from its solutions in other acids. Phosphoric acid, applied on the dry powder, evolved gas and dissolved a small amount. The solution had a sour, contracting (astringent) taste, and became gelatinous under evaporation. If the phosphoric acid was saturated with the earth, the substance was insoluble in water. After drying, this saturated compound weighed more than twice as much as the earth. It did not melt in the blow-pipe and gave with borax a yellowish glass, becoming opaque in the outer flame.

The reaction with *boric acid* took place through double decomposition. A borax solution was added to a solution of the earth in another acid (one part of earth in 7 parts diluted sulphuric acid: 3 parts of water and 1 part of concentrated acid). The result was mixed in 100 parts water. Into this, the borax solution was gradually added. This gave a temporary opalescence disappearing again in acid solution. Later a white *["gelb/hvit"!]* precipitate appeared, ceasing after 5 parts of borax had been added, the solution still remaining slightly acid by a litmus test. The obtained precipitate, weighing about one-half of the dissolved earth, melted in blow-pipe to a milky pearl. With added sodium ammonium hydrogen phosphate the pearl became white and opaque. With borax a half-clear glass was obtained. After extended blowing it became completely clear.

Under slow evaporation the clear solution first gave a number of clear, needle-shaped crystals, then some larger octahedral clear crystals and after that clear boric acid and Glauber's salt [$Na_2SO_4 . 10H_2O$]. The needle-shaped crystals had limited solubility in water, little taste and disintegrated when heated, melting at high temperatures to a white enamel which finally clarified into a glass pearl. The octahedral crystals were more easily soluble, had a sweet taste and left a feeling of heat on the tongue. They contained plenty of water of crystallization, causing foaming in the blow-pipe. At higher temperatures the salt became white and opaque, melting afterwards to a clear glass.

Oxalic acid precipitated the earth from its solution in all other acids, unless an excess of that acid was present. Thus the earth was entirely precipitated from its solution in sulphuric acid when oxalic acid, saturated with potassium hydroxide, was added. The precipitate was a white powder that was carbonated in the blow-pipe, burned with a fiery flame, then becoming white and behaving like the original earth. A part of this precipitate was calcined in a crucible whereby it lost half of its original weight. The heated powder, dissolved in sulphuric acid, gave by slow evaporation clear, irregular crystals. When all superfluous sulphuric acid had been dried away, and the dry residue dissolved in water, renewed evaporation gave sharp, rhombic crystals, having a sweet taste and contracting the tongue. These crystals were stable in air. In the blow-pipe they disintegrated, forming thin sheets that did not melt alone at high temperatures and gave a milky glass with borax . One part of these clear crystals required 22 parts of water for dissolution at normal temperature. Because such crystals were not obtained when only the earth was dissolved in sulphuric acid, it appears that some part of the organic acid had survived the calcination.

The earth was similarly precipitated by a saturated alkaline solution of oxalic acid from its solutions in hydrochloric acid. As, however, this solution had a strong excess of acid, complete precipitation of the earth was not obtained. After evaporation the mother liquor gave under evaporation clear, rhombohedral crystals resembling gypsum and consisting of parallel, rhombohedral sheets. These were dissolved in the blow-pipe in their own water of crystallization. At higher temperatures they were carbonated and transformed into porous carbon, melting finally into a white, opaque pearl, having a caustic, alkaline taste. Hence this salt seems to consist of the earth, combined with oxalic acid and potassium hydroxide. It dissolved quite easily in warm water but required, at average temperatures, 45 parts water to one part salt. Addition of *tartaric acid* to a solution of the earth in hydrochloric acid gave no precipitate. If, however, the extra acid had been neutralized by alkali and the added tartaric acid was likewise saturated with potash, a large white precipitate resulted. It seemed to be more soluble in pure water than the oxalic compound.

Acetic acid dissolved the earth easily, with evolution of gas.

Johan Gadolin (Courtesy of the National Museum of Finland)

Conclusion

"These properties suggest that this earth has similarities with alum and with CaO, but also dissimilarities from both of these, as well as other known earths, whence it should be included among the simple earths, unless the experiments lead to a conclusion that it is composed of several substances. I do not yet dare to claim such a discovery, both because my limited supply of the black mineral has not permitted all desired experiments and because science would benefit more, if the new earths recently described by chemists could be separated into simpler constituents, than in the case that their number is further increased" (Gadolin, 1794, 1796).

1.4. Gadolin's letter to Wilcke

The manuscript of Gadolin (1794) was mailed to Professor J. C. Wilke. In the covering letter, dated 22 May 1794 (in Swedish, reprinted on p. xci of Hjelt and Tigerstedt (1910)), Gadolin thanks Cap. Arrhenius for providing a few years ago the sample of the black, shiny stone, found among the red feldspar in the Ytterby mine.

"Had I had more of the stone I would not have considered my experiments finished".

"It is not without great trepidation, I dare speak of a new earth, because they are right now becoming far too numerous, unless one rapidly finds a way to analyse them further. However it seems to me that mine should have some similarity with Klaproth's zirconia"...

"Should the Ytterby earth be identical with zirconia, much would be gained, for it seems to me rather fatal if each of the new earths should only be found at one site or in one mineral".

As a curiosity we note the Editor's foot-note on p. 313 of Gadolin (1796): The author's submission letter was dated 3 July 1794 but reached Crell on 26 March 1796, perhaps because the letter had apparently been given to a traveller. "Meanwhile, the paper has lost nothing but, would it be written now, it should contain some clarifications due to Prof. Klaproth's excellent contributions".

1.5. The first confirmation by Ekeberg

The first confirmation of Gadolin's analysis was published by Ekeberg (1797), who had obtained for this purpose from Captain Arrhenius a larger sample without feldspar. This diminished the amount of silica and aluminium oxide obtained. His result was:

Silica	25 parts
Iron oxide	18 parts
Aluminium oxide	4½ parts
The new earth	<u>47½ parts</u>
	95 parts
Missing	5 parts.

[In retrospect, the mineral "gadolinite" is $FeBe_2Y_2Si_2O_{10}$. The beryllium was confused with aluminium.]

The properties of the new earth fully agreed with those, reported by Gadolin. Ekeberg further mentions the sweet taste, "almost like that of lead compounds, but not so repulsive [*äcklig*] but rather contracting [*sträv och adstringent*]. The acetate was to my taste just as sweet as lead sugar [lead (II) acetate]."
Beautiful, air-stable crystals of the sulphate and acetate were obtained and described in detail. Solubility in arsenic acid and the slight solubility of the arsenate were discussed. Ekeberg gives a clear summary of the differences between the new earth and the previous ones. He proposed the Swedish and Latin names of "Ytterjord" and "Yttria", respectively, for the earth and "Yttersten" for the mineral. Finally he suggests that yttria, whose solutions have such an original taste, could also eventually be medically potent.

1.6. Acknowledgement

We thank Margareta Jorpes-Friman, Åbo Akademi Library, for providing us with copies of the original articles.

1.7. References

Asimov, I., 1976, *Biographical Encyclopedia of Science and Technology*, Avon Books, New York, p. 223.

Ekeberg, A.G., 1797, "Ytterligare undersökningar av den svarta stenarten från Ytterby och den däri fundna egna jord", *Kungl. Svenska Vetenskapsak. Handl.* 156-164.

Enkvist, T., 1972, *The History of Chemistry in Finland* 1828-1918, Soc. Sci. Fennica, Helsinki, 161 p.

Gadolin, J., 1792, "De natura metallorum", as reprinted in Hjelt and Tigerstedt (1910).

Gadolin, J., 1794, "Undersökning av en svart tung stenart ifrån Ytterby stenbrott i Roslagen", *Kungl. Svenska Vetenskapsak. Handl.*, 137-155.

Gadolin, J., 1796, "Von einer schwarzen, schweren Steinart aus Ytterby Steinbruch in Roslagen in Schweden", *Crells Ann.*, 313-329.

Geijer, B.R., 1788, (letter to editor, without title), *Crells Ann.*, 229-230.

Hägg, G., 1963, *Allmän och oorganisk kemi*, Almqvist & Wiksell, Stockholm.

What did Johan Gadolin do?

Hjelt, E. and Tigerstedt, R., 1910, *Johan Gadolin 1760-1852 in Memoriam*, Acta Soc. Sci. Fenn. **39**, 96 + 287 p.

Holleman, A.F. and Wiberg, E., 1976, *Lehrbuch der anorganischen Chemie*, Walter de Gruyter, Berlin and New York.

Klinge, M., Knapes, R., Leikola, A. and Strömberg, J., 1987, *The Royal Academy at Åbo 1640-1808* (Finnish and Swedish editions), Otava, Helsinki.

Mäkitie, O., 1982, "Johan Gadolin ja harvinaiset maametallit" (J.G. and the rare earths, in Finnish), *Kemia-Kemi*, **9**, 85-88.

Pearce, J., 1987, *Gardner's Chemical Synonyms and Trade Names*, 9th Ed., Gower Technical Press, Aldershot, Hants.

Pyykkö, P. and Orama, O., 1988, "Johan Gadolin's 1788 paper mentioning the several oxidation states of tin and their disproportionation reaction", *New J. Chem.* **12**, 881-883. (Misprinted portrait printed correctly ibid. **13** (3), 269.)

Römpp, H., 1966, *Chemie Lexicon*, Franckh'sche Verlagshandlung, Stuttgart.

Sjöblom, L., 1960a, "Johan Gadolin 1760-1960", *Svensk Kemisk Tidskrift* **72**, 748-756.

Sjöblom, L., 1960b, "Johan Gadolin - ett tvåhundraårsminne", *Finska Kemists.Medd.* **69**, 54-65.

Tigerstedt, R., 1899, *Kemiens Studium vid Åbo Universitet* , Åbo Universitets Lärdomshistoria, Vol. 8, Svenska Litteratursällsk. i Finland, Skrifter **42**.

Toivanen, P., 1980, *Johan Gadolin ja Aineen Rakenne* (Johan Gadolin and the Structure of Matter, in Finnish), Report HU-P-D18, Department of Physics, University of Helsinki, 196+11 p.

Appendix 1- Historical nomenclature

Original Name Swedish/German	Page Swedish/German	Present English Name	Formula	Reference
Acetosellsyra/Acetosellsäure	153/327	Oxalic acid	$C_2H_2O_4$	d
Alkali/Laugensalz	142/318	Sodium or potassium carbonate	Na_2CO_3,K_2CO_3	a
Alkali aeratum/Luftgesäuertes Alkali	151/325	Sodium or potassium carbonate		b
Alkali minerale/Mineralalkali	143/318	Sodium carbonate	Na_2CO_3	c
Alun/Alaun	143/318	Alum	$K_2Al_2(SO_4)_4\cdot 24H_2O$	c
Alunjord/Alaunerde	143/318	Aluminium oxide	Al_2O_3	b
Blodlut/Blutlauge	145/320	Potassium ferrocyanide	$K_4[Fe(CN)_6]$	d
Caustik pottaska/Kausticher Laugensalz	146/321	Potassium hydroxide	KOH	b
Caustict flygtigt alkali/ Kausticher flüchtiger Laugensalz	144/319	Ammonia (concentrated)	NH_3	b
Flyktigt alkali/Flüchtiges Alkali/ Volatile alkali	144/319	Ammonia	NH_3	b
Flusspat/Flusspat	141/317	Calcium fluoride	CaF_2	d
Fältspat/Feldspat	137/313	Potassium aluminium silicate	$KAlSi_3O_8$	d
Järnkalk/Eisenkalk	146/321	Iron oxide	Fe_2O_3	b
Kiseljord/Kieselerde	144/319	Silica	SiO_2	b
Luftsyra/Luftsäure	151/325			
Sal microcosmicus/Harnsalz	150/324	Sodium ammonium hydrogen phosphate	$Na(NH_4)HPO_4\cdot 4H_2O$	b
Sal microcosmicus/Weinsteinsalz	139/315	Potassium hydrogen tartrate	$KC_4O_6H_5$	b
Sockersyra/Zuckersäure	153/327	Oxalic acid	$C_2H_2O_4$	
Vegetabiliskt alkali/Gewächsalkali	145/320	Potassium hydroxide	KOH	b

References:
a. Holleman-Wiberg (1976)
b. Pearce (1987)
c. Häg g (1963)
d. Römpp (1966)

Appendix 2 - The analysis schemes used

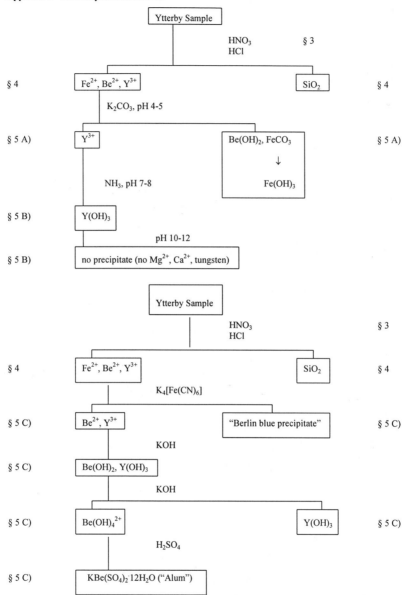

CHAPTER 2

THE DISCOVERY OF CERIUM - A FASCINATING STORY

JAN TROFAST
Centre for the History of Science
The Royal Swedish Adademy of Science
Stockholm, Sweden

2.1. Introduction

The discovery of a new element is not an isolated event. Several small consecutive observations often lead to new presentations of a problem. When the time is ripe and the different observations combine with new knowledge, all the bits and pieces fit together and a beautiful new discovery is made.

The discovery of cerium is a brilliant example of the activities which took place in chemical laboratories in Sweden at the beginning of the 19th century. It illuminates the discussion and the choice of problems at that time, the uncertainty of the analytical techniques and the difficulties in the interpretation of the results. What directed the chemist's choice of a problem? The tradition within mining established almost immediately the field of mineral analysis, although it had become very popular to investigate the effects of the recently described voltaic pile. The high activity surrounding pneumatic chemistry around the 1770s and 1780s had decreased. A reason for the analysis of a certain mineral could be a divergent physical property such as density or crystal form, which could not be explained by the knowledge of that time. Seldom a practical problem influenced the direction of mineral analysis. Yet scientists often discussed solutions to technical problems associated with the handling of slag, the construction of furnaces etc, and several journeys were made by Swedish mining engineers to study mining mechanics and the iron industry abroad.

Axel Fredrik Cronstedt, chemist and assayer, had described the discovery of two minerals with an exceptionally high density in the Transactions of the Swedish Academy of Sciences in 1751.[1] They were named tungsten ("heavy stone") and had been found at Bispberg near Säter and Nya Bastnäs or St Görans mine in the neighbourhood of Riddarhyttan. After preliminary experiments, Cronstedt suggested that the mineral from Riddarhyttan should be tested for an assumed content of zinc and

13

C. H. Evans (ed.), Episodes from the History of the Rare Earth Elements, 13–36.
© 1996 Kluwer Academic Publishers. Printed in the Netherlands.

he remarked that there were copper- and bismuth-containing minerals interspersed. The interspersion of other minerals can probably explain the differences in density between different specimens.

Cronstedt introduced the tungsten from Bispberg under the title of iron in his 1758 anonymously published "*Försök till en Mineralogie eller Mineralrikets upställning*" (An Essay towards a System of Mineralogy). It was described as a calcined iron bound to another unknown earth. He called the mineral "*Ferrum calciforme terra quadam incognita intime mixtum*" *or just* "*tungsten*". It was hard to reduce and melt.[2] Cronstedt was close to the truth in his analyses, but it was Carl Wilhelm Scheele who further investigated the composition of the tungsten. Scheele isolated and studied the properties of the metal compounds by use of different precipitation and redox reactions, and documented the different series of salts. This procedure became a general technique, which was frequently used by chemists to prove definitely the existence of new elements. In the spring of 1781 Scheele could describe his investigations of the hitherto unknown acid - the tungstic acid (wolframic acid, WO_3 - H_2WO_4) - through the analysis of the tungsten. The mineral is now called scheelite (calcium tungstate, $CaWO_4$).[3]

Torbern Bergman confirmed the result,[4] but it was the Spaniard Juan José d'Elhuyar who, together with his brother Fausto, received the honour for having been able to reduce the tungstic acid to the new element tungsten (wolfram). When they heated a mixture of tungstic acid with charcoal for an hour and a half the metal was formed. The name of scheelium had also been proposed but Jacob Berzelius thought, later on, it had better keep the name of wolfram.[5] During some months in the spring of 1782 d'Elhuyar studied at Bergman's laboratory in Uppsala and thus had a reason to make these analyses. Scheele had isolated molybdic acid earlier. Although this acid is chemically closely related to wolframic acid, a different analytical technique was needed. Molybdenite (MoS_2) is oxidized, while this is not the case with tungsten; barium molybdate is soluble in water while barium tungstate is not, and molybdic acid (MoO_3 - H_2MoO_4) is volatile when strongly heated unlike tungstic acid. The analytical technique and, especially, the knowledge of solution chemistry (wet analysis) were improved.

Already in the middle of the 1780s the German chemist M H Klaproth worked with minerals from Cornwall in England, and obtained precipitates with the same properties as Scheele had reported from his analyses of the mineral from Bispberg.[6]

In January 1782 the mere 15-year-old Wilhelm Hising (later Hisinger after being raised to the nobility), who studied with Bergman in Uppsala, sent the reddish mineral from Bastnäs to Scheele for analysis. Scheele concluded that Bastnäs tungsten did not contain any tungstic acid but only silica, much alumina and traces of iron.[7] One circumstance which might explain Scheele's incomplete analysis was the fact that he was, at the time, in the middle of rebuilding his laboratory and therefore had less time for chemical analysis. It is surprising that he should hand over a seemingly incomplete result of the analyses. Another circumstance was the fact that the analyses were extremely difficult.

D'Elhuyar also analysed the tungsten from Bastnäs and got, after analysing 1 centner (1 centner assay balance = 3.32 g) of this mineral, nothing but 24 parts of iron, 22 parts of silica and, as to the rest, just lime.[8]

The difference between Scheele's and d'Elhuyar's results can be due to the fact that d'Elhuyar precipitated the earths with oxalic acid and then believed he had got calcium oxalate, while Scheele used ammonia and believed that the origin of the precipitate was alumina.

2.2. Gadolinite

Some years later (1787) the young artillery officer Carl Axel Arrhenius reported that he had found a strange black mineral at the felspar quarry in Ytterby near Stockholm. Sven Rinman classified it under the heading Pechstein in his famous Bergwerks lexicon (1789).[9] Bengt Reinholm Geijer found its density to be higher than that of earlier known earths.[10] The first chemical analysis was made by Johan Gadolin in Åbo (Turku). He found an unknown white earth, whose behaviour before the blow pipe and with acids reminded him in certain aspects of alumina, but in others, of lime. The results of the analysis were 31 parts of silica, 19 parts of alumina, 12 parts of calcined iron and 38 parts of an unknown earth.[11] Anders Gustaf Ekeberg made a new investigation of the mineral and produced several new compounds following Scheele's method. Ekeberg's analyses confirmed that it was a new earth and it was given the name of yttria after the place where the mineral was found for the first time.[12] The mineral was later named gadolinite after Gadolin. The result of his analyses was 23 parts of silica, 55.5 parts of yttria, 4.5 parts of glucina, 16.5 parts of oxide of iron and 0.5 part of volatile matter.[13] Ekeberg did not succeed in finding the cerium oxides or the chemically closely related elements (the lanthanides) in gadolinite. The analyses of Gadolin and Ekeberg are described in detail in the previous chapter of this volume.

Interest in the newly found earth increased among chemists. Gadolinite is a very rare mineral and is found primarily in Ytterby and near Broddbo, Finbo and Kårarvet in the neighbourhood of Falun in Sweden. Several more or less careful analyses of the gadolinites from various places were made as time went on. The results show the difficulty in unequivocally and correctly describing the chemical formula of a mineral. Selected specimens often contained mixtures of other minerals which sometimes made it difficult to establish the chemical composition with certainty. Furthermore it is astonishing to observe the fast development of the methods and the improvement of the accuracy of the analyses. Much of this development within analytical chemistry can be ascribed to Jacob Berzelius.

Wilhelm Hisinger (1766-1852) Jernkonstoret, Sweden

2.3. Bastnäs tungsten (Cerite)

2.3.1. THE QUALITATIVE ANALYSES

Ironmaster Wilhelm Hisinger speculated whether yttria could be found in the minerals round Skinnskatteberg and Falun. The first analyses were promising and the discussions were lively. When young Jacob Berzelius in the spring of 1803 visited him in Västmanland, it was natural to discuss the different minerals with gadolinite as a starting point. On this occasion they also discussed the earlier analyses of the Bastnäs tungsten which, in Scheele's opinion, did not indicate tungstic acid. Then the question was whether the high density of the mineral was due to yttria. In their preliminary experiments they found an element which had several characteristics in common with yttria, but which differed in others. After many comparative experiments they drew the conclusion that a completely new element had to be involved. The investigations continued in Stockholm in the winter of 1803-4.

The introductory experiments seemed to have been completed at about the turn of the year 1804. Berzelius, in the usual way, made notes of the results of the experiments with Bastnäs tungsten.[14] At this time he used the symbol system which was published in 1787 by the Frenchmen J H Hassenfratz and P A Adet. It is most notable that, already at an early stage of his investigations, he was sure that he had found the oxide of a new metal. The new element got the preliminary name of *bastium* - the name derived from the place of discovery of the metal. The metal could occur in two stages of oxidation (protoxide and peroxide) and thus be the origin of two series of salts, one colourless and the other coloured yellow or red. The name of the metal was soon changed to *cerium* and the name of the mineral to *cerite* after the planet Ceres which had been discovered in 1801 by the Italian astronomer Giuseppe Piazzi.

As usual Hisinger notified assessor Johan Gottlieb Gahn in Falun of the results. The description was so detailed that Gahn could get an insight into the problem and judge if any mistakes had been made in the analyses.

> Since I came to town Berzelius has been occupied with the prolonged analysis of the Bastnäs heavy mineral, whose quantitative analysis is now well started. Since we earlier had found great quantities of a metallic oxide, which seemed to be new and did not correspond to any earlier known metallic oxide and whose properties I briefly would like to submit to your scrutiny and which has recently been named Cerium.
> The easiest way to get it pure, though not without a loss, is to let strongly heated tungsten be dissolved in nitric acid, which should not contain sulphuric acid. Ammonia is added and the precipitate formed is dissolved in muriatic or acetic acid and then treated with a solution of sulphate of potash, both of them saturated before. Then the result is a white or yellowish precipitate consisting of sulphuric acid, potash and this oxide.

Besides Berzelius has gone through its relation to almost every
acid and precipitation agent which would be too detailed to write
about in a letter but which all seem to confirm the different
character of the oxide compared to the known oxides; we will
shortly try a reduction experiment at Mr Hjelm's.[15]

In the property left by Gahn there was a manuscript "*Närmare bestämmande af den i Bastnäs Tungsten innehållna Metall-oxidens egenskaper*" (An accurate determination of the properties of the metallic oxide found in Bastnas Tungsten) and it seems to have been written by Berzelius. The manuscript is not dated, but it is plausible that it was written at the end of January 1804. It differs somewhat from the published paper on cerium. There is, among other things, no physical description of the mineral. Having reported a great many experiments Berzelius ends with a precautionary measure:

All these properties taken together are not to be found in any
other specimen, that we have found described before, which, for
the time being, has given us reason to believe that cerium is a
metal of its own. It would please us if our report later on could be
confirmed by the Masters of the Art of Analysis and we would
also without displeasure accept a correction if we, in our
investigations, should have made any mistakes.[16]

Hisinger also continued his correspondence with Gahn and on January 30, 1804, he wrote:

With the deepest gratitude I am sending the oxide for reduction
experiments. It is almost completely purified from sulphuric acid
and iron. I await the report on their success with great interest.
The experiment at Mr Hjelm's left only a black powder, which
must, to some extent, have been reduced since it produced
hydrogen gas with acids. I also take the liberty to send an essay
on the properties of the oxide which is practically correct. The
final one is being translated into German for Journal der
Chemie.[17]

Hisinger to Gahn on February 14, 1804:

The paper on the oxide is sent off, wherein the Assessor's kind
promises of the melting experiments were mentioned and if
permitted will be a continuation thereof. Berzelius has promised
to let me shortly send you more of the oxide, as I believe it would
be worth trying to melt it with silver and copper and with some
reducing fluxes.[18]

The reduction of a metal oxide was made through mixing the finely pulverized metal oxide with the reducing agent (often charcoal) and a flux. The reduction was performed under strong heat in a covered crucible. Berzelius had, possibly in connection with sending the above mentioned act, sent a letter to Gahn, which reads as follows:

May I express my utmost gratitude to you, sir, for the kindness you have shown in devoting time to the reduction of the cerium oxide. This will probably be difficult and perhaps the only possible way to do it is the way you, sir, have started it. At Mr Hjelm's, Keeper of the Mint, Hon. Mr Hisinger and I tried to reduce it on its own. We received a black powder which was similar to the black masses, that you, sir, sent us; it dissolved in muriatic acid first with evolution of hepatic air and then with continuous evolution of gas and a smell of hydrogen gas just like when iron is dissolved.

Maybe this metal in a reduced state combines with carbon and consists of this black refractory powder; at least the dissolution in muriatic acid left much carbon? In this case all efforts to isolate a pure metal without a mixture of other substances will probably be in vain.[19]

Jacob Berzelius (1770-1848) Painted by P.O. Finnstrom 1979. Private

Gahn thanked him for the parcel by return.

> Thank you ever so much for your letter and for the new oxide you
> sent me, which I am curious to analyse further, as soon as I have
> got a few urgent matters off my hands, perhaps in a fortnight. This
> carbonate of cerium is completely different from the other two, it
> is all white and dissolves easily with effervescence in sulphuric,
> nitric and muriatic acids. Dissolved in a test glass by the uric salt,
> if you do not take too big a portion, it sometimes leaves in weak
> but fuliginous reduction flame spots on the glass, which have a
> metallic shine somewhat darker than iron: and in a strong
> reduction flame the glass is coloured slightly blue, especially if so
> much is added that it, when cooling off, becomes a little turbid. It
> is possible that the heating temperature cannot be too strong when
> reducing the oxide. The gentlemen in Stockholm, I am sure, had a
> weaker flame than we here and still the gentlemen had a more
> distinct deoxidization through a distinct smell of hydrogen gas
> and so on. Was the oxide mixed with nothing else but water in
> this experiment? And placed in a prepared crucible covered with
> coal cinder.
>   Why does the purest carbonate of cerium, that you sent me first,
> not dissolve in acids but to a trifling extent? Is this due to too
> great an oxidation? It cannot contain so much sulphuric acid, that
> this can be the cause, can it? Could it have changed during the
> transportation? I return some of it for an examination by Hon. Mr
> Hisinger.
>   Why does the brown sulphate of cerium which is saturated and
> insoluble, not dissolve in fresh dilute sulphuric acid as it dissolves
> in it from the beginning with the same degree of oxidation? And
> how can it be utilized and the oxide be separated from the acid?
> Also with potash and strong heat? You, gentlemen, would also
> give us the analysis of the mineral and tell us the easiest way to
> educe the calcinated metal.[20]

On receipt of the sample Gahn had immediately tested the reactions of the mineral before the blow pipe, and as soon as he had the opportunity he had gone to his laboratory for further studies by chemical methods. His test records, which are still preserved, follow Berzelius's paper.[21] Unlike Berzelius, Gahn used the symbol system which had been developed by Torbern Bergman. The effects of cerium oxide very much resemble those of iron oxide, particularly if the cerium is combined with silica; the protoxides of iron and cerium do not behave alike with respect to the fluxes. Nevertheless, when in simultaneous combination with silica, as they usually are, the presence of cerium oxide cannot be ascertained by the blow pipe. Heated alone on platina the protoxide passes to the state of peroxide. On charcoal the peroxide does not alter.

Johan Gottlieb Gahn

The impatient and restless Berzelius examined Gahn's annotations and shortly sent off another letter:

Dear Sir,

Thank you very much for your interesting letter. The experiments with the Ceroxide and the blow pipe were striking and give great support to your surmise that a reduction probably will succeed in a weaker flame. In the experiment at Mr Hjelm's we had the oxide in a carbon crucible which was tightly closed by a stopper of carbon. The oxide was earlier burnt with linseed oil à la Klaproth and it was put there in the shape of small pieces. After the experiment it had fallen apart and become a rather fine powder. The carbonate of cerium from last time was not as pure as I thought. It was produced from sulphate of cerium melted with 3 times as much crystallized carbonate of potash in a platinum crucible. Thus there was reason to believe that it was pure. But the decomposition was incomplete, only a small part is carbonated and the rest is the persistent combination of the oxide and sulphuric acid.

Why the brown sulphuric acid-containing oxide did not dissolve in dilute sulphuric acid is hard to explain. If you boil it with an acid diluted with 2 to 3 parts of water, it is even harder. I have in these experiments always first boiled it with strong sulphuric acid to get it supersaturated with acid and then poured away a part of the above mentioned acid, mixed the rest with water and then everything has dissolved. The acid that was poured away contained very little dissolved material, probably because of lack of water.

You, sir, advise us to give an analysis of the stone. We have realized the necessity for an exhaustive paper, but neither Ironmaster Hisinger nor I have enough experience of the analytical way of working or good enough measuring devices to make such an accurate analysis that men like Ekeberg and Klaproth would not have much to comment on our work. Therefore we have only reported the approximate proportion that we found in our experiments to educe cerium without any pretentions on giving the accurate composition of the mineral. We have found 0.23 silica, 0.05 carbonic lime, 0.22 iron and 0.50 Cerium. There is probably no loss, but these 50 cerium surely contain 10 percent more acid than the mineral. When analysing minerals I have an objectionable habit of not being patient enough to be accurate. To know whether a stone contains three to four hundredths more of this or that substance has never a value which can answer to the sacrifices I make out of my curiosity and the pain I expose my natural little patience to when I for several days have to wait for a precipitate to have completely deposited or when I for ever and ever must rinse a filtrate until the rinsing water has no taste and shows no reaction. If the result of the analysis has enough interest for me then I can sit half the day in peace and quiet watching it slowly proceed. Therefore it happens that some of my experiments are done with embarrassing conscientiousness and others with much carelessness depending on the result I am expecting.[22]

Gahn's notes on the experiments are not dated but one can presume that Berzelius tried to wait for Gahn's results as long as possible before he sent off his paper, which was written in German, to A F Gehlen to be published in his *Neues allgemeines Journal der Chemie*. The Swedish version had been written on January 28, 1804.

Berzelius also had already dutifully told his former teacher Professor Johan Afzelius in Uppsala, about the new discovery on January 12, 1804. Meanwhile there was a busy correspondence between Gahn in Falun, Afzelius and Ekeberg in Uppsala and Berzelius and Hisinger in Stockholm regarding possible mistakes in the analyses.

It would be interesting to see how Afzelius and Ekeberg received the news about the discovery of a new element. Afzelius had let Ekeberg make a few analyses of the specimen he had received, but otherwise these chemists seem mostly to have speculated on the inexperienced Berzelius and, which would be proved later on, his excellent ability to make chemical analyses.

Berzelius wrote about this to Gahn:

> Prof. Afzelius has written to me and told me that they had started to investigate Bastnäs tungsten in Uppsala and that Ekeberg has found Yttria in it and an unknown substance, of which he wanted to inform me, to avoid mistakes if Hisinger and I were in a hurry with our experiments. It would annoy me immensely if we were mistaken [Berzelius first wrote: "if I was mistaken"] about Cerium, if this is a mixture between yttria and iron or such like. But I am conceited enough to think that Ekeberg, when he told Afzelius that there is yttria in the specimen, did not carefully enough investigate his yttria, but, just as we did in the beginning, judged it by its sweet taste, its solubility in acetic acid into a crystallizable salt and its sparing solubility in sulphuric acid. Today I will send them a portion of the oxide for investigation. Let us see if they get yttria out of it and if so, I do not fear anything.[23]

The following letter is from Berzelius to Afzelius March 7, 1804 where he further presents the results he has achieved from his experiments:

> I wish to express my humble thanks for the information on Mr Ekeberg's investigation of the Bastnäs tungsten or as you, Professor, called it, the Bispberg tungsten. It will be a pleasure to examine the results of his experiments, which are performed with more accuracy than ours, as far as the composition of the mineral is concerned. I have not been able to find yttria but that is perhaps due to the fact that I only know its properties from what I have been told; but there must be a rather small quantity in it.[24]

Hisinger to Gahn April 26, 1804:

> I hear from Professor Afzelius that he regards this [the cerite] to
> be a manganiferous yttria earth mixed with some glucina; these
> remarks however are said to be made without any experiments
> and just after reading the essay.[25]

Berzelius got rather annoyed with the lightness with which Afzelius dismissed the new results and he was not slow to reply to the criticism with scientific cogency:

> Gentlemen, your opinions about our new metallic oxide have
> really made me puzzled the more so as I have seen so much proof
> of even the most experienced chemists being wrong. I have
> however great difficulties in convincing myself that manganese is
> the cause of the metallic nature of this substance. Can manganese
> in combination with earths change with regards to its properties so
> that it cannot be dissolved by caustic alkali even at a high
> temperature? Can it in this combination absorb and evolve
> oxygen without the combination being destroyed? Can it,
> furthermore, be held in a combination in such a large quantity,
> which can be observed from the formation of the oxidized
> muriatic acid gas when the oxide is dissolved in acids, that it
> exists in the cerium oxide without differing, in any way, from the
> earth, with which it is combined? If all this should happen, then
> our delusion is excusable.
> There is no single earth and no metallic oxide which has been
> discovered lately that I have not thought it could be and with
> which I have not compared it. In particular I have compared it
> with the manganese oxide; I got it from manganese dioxide and
> always used the same method to purify it from iron as I did with
> cerium but the differences have always been rather substantial.
> Especially I refer to the yellow colour which this oxide shows in
> the highest degree of oxidation with sulphuric acid and muriatic
> acid and so on. I am sure that I mentioned that the first way of
> educing cerium was to precipitate its solution in nitric acid with
> sulphate of potash. Not one of these three substances, which
> according to your opinion, gentlemen, is the cerium oxide, is
> known to have this property. The manganese has it not, I am sure.
> If yttria should produce such a salt, the manganese ought not to be
> separated here; or when, in accordance with your theory, it is not
> separated, it would not, at any rate, when burnt with potash, with
> saltpetre, with tartar appear in one of its oxidation stages?[26]

In Uppsala rumours also spread that Berzelius and Hisinger had drawn too hasty conclusions about their discovery. On direct questioning Afzelius had to confess that he had only expressed a guess about the composition of the mineral based on its density. In spite of the disputes between Berzelius and Afzelius regarding the analyses of the cerium mineral, they do not seem to have caused any bitterness in either party. Berzelius was invited to stay with Afzelius when he was in Uppsala because of the conferment of his doctor's degree (May 24, 1804).

While disputing with Afzelius, Berzelius had received an answer from Gehlen that the paper he had sent, had been approved for immediate publication; but, in the third issue, which was being printed, he had a report with the discovery of a new earth from the same mineral written by the German chemist M H Klaproth in Berlin.[27] What a disappointment for the mere 24-year-old man from Väversunda! He could come out of the struggle with Afzelius victorious, but how would the world around him evaluate his achievement? This especially influenced Hisinger's behaviour. Again it was Gahn who was the first to be informed:

> I think I should inform you, sir, about the misfortune, which the discovery of the new metal has suffered. Ironmaster Lindbom, to whom I mentioned the discovery, has answered from Paris that Vauquelin has shown a sample of a new earth from Bastnäs tungsten, which he at the end of February received from Klaproth and which he called Ocharit or ochrorit because of the colour which resembled iron ochre. Consequently it must be the same substance that we for many reasons believe to be an oxide. I do not exactly remember the time I sent the paper to the publisher, Mr Fröhlich; but it must have been at the beginning of February, at the latest. - Fröhlich has not answered but I have now written to ask whether it has arrived, which cannot be doubted. If this really is a competition, it will be a pleasure to see which side Klaproth will support and which will be apparent in the next issue of Ch. Journ. for March, which soon will arrive. I will however send a copy to Lindbom to do with whatever he finds suitable. When someone remarked to Vauquelin on Klaproth's new earth that the colour seemed more to characterize an oxide, he had become angry and answered that Klaproth no doubt could tell an earth from a metallic oxide![28]

Jacob Berzelius

A clear-cut definition of an earth did not exist in the 18th century. Earth denoted a mineralogical substance which lacked taste, was insoluble in water, was incombustible and which was the basis of the minerals. Silica and alumina, for instance, could be regarded as oxides or hydrated oxides. Lime, magnesia and barytes denoted at first the carbonates, but later the oxides. At the end of the century an earth was described as an oxide which was difficult to reduce, difficult to dissolve and difficult to volatilise. At the end of the century there was a clear-cut difference between an earth and a metallic oxide.

What analytical technique had Hisinger and Berzelius used to trace the new metallic oxide? After having removed silica and iron through treatment with concentrated acids, the brownish powder was treated with concentrated acetic acid, while being heated. As it cooled, crystals appeared, which were suspected of being acetate of yttria. The part of the precipitate which was dissolved in the ethanol was treated with concentrated ammonia and, after filtration, calcium carbonate was precipitated with ammonium carbonate. The crystals which were formed with the acetic acid thus consisted of chalk and an unknown substance which was brick red after heating. This was dissolved in hydrochloric acid with the evolution of chlorine, which Berzelius interpreted as if the powder contained oxygen. At that time chlorine was considered to be oxygenated hydrochloric acid! He drew the conclusion that it had to be a metallic oxide. Then it was a question of excluding all known metallic oxides through precipitation reactions and of characterizing the salts of the new metal. Salts of the following acids were identified: nitric acid, hydrochloric acid, phosphoric acid, arsenic acid, molybdic acid, oxalic acid, tartaric acid, carbonic acid, benzoic acid, citric acid, acetic acid, succinic acid, gallic acid and prussic acid (hydrogen cyanide). Together with the blow pipe analysis of the metallic oxide, these experiments fully satisfied the criteria of a new element. The results might at first sight seem rather inexact, but then one must bear in mind the demand for high experimental skill to arrive at the necessary judgment. The iron oxide in the cerite, for instance, gives almost the same colour spectrum when melted before the blow pipe, together with different fluxes, as the ceroxide.[29]

Because of the reports from Gehlen and Lindbom in Paris, Hisinger had had 50 copies of the nowadays very rare publication "Cerium, a new metal, found in Bastnäs Tungsten" printed in May 1804 at his own expense to be distributed among his scientific friends.

The article on cerium, which was published in the fourth number of Gehlen´s journal, attracted great attention among the chemists in Europe. Shortly it was published in French in *Annales de Chimie* and in English both in Nicholson's Journal of Natural Philosophy and in Tilloch's Philosophical Magazine.[30] Through the almost simultaneous publication of the discovery by Klaproth, priority disputes seemed to be near. Disappointment was great on both sides and false rumours spread.

Hisinger and Berzelius had their suspicions about Klaproth having read their paper at Gehlen's but those seem to be completely unfounded speculations.

Klaproth announced his discovery in a letter to L N Vauquelin, who had it read in the French Academy of Sciences. An account of this was published in *Annales de Chimie* (an XII)[31] and almost immediately there were translations in English journals. Vauquelin had learnt about the Swedes' discovery through Lindbom's translation in Paris. The latter had immediately translated into French the article which had been sent to him by Hisinger. Because of this procedure Hisinger and Berzelius called attention to their priority. Vauquelin already wrote a letter on June 1, 1804 addressed to: *Monsieur Hisenger, Chemiste et mineralogiste celebre à Stockholm, Suede.* The contents are of principal importance to the priority dispute between Klaproth on one side and Hisinger and Berzelius on the other. From this letter it is evident that Hisinger and Berzelius thought that they had found a new metal while Klaproth speaks of a new earth!

> *Ce chimiste [Klaproth] la regarde comme une terre qui suivant lui fait le passage entre les Substance de ce genre, et les oxides metalliques, à cause de sa pesateur et de la couleur qu'elle prend au feu. C'est une idée qu'il a concue, il y a deja longtemps, et qu'il a exprimée au Sujet de l'Yttria. Vous, Messieurs, au contraire vous considerés cette Substance comme un métal particulier, et je partage volontiers votre opinion à cet égard. J'ai meme, avant depuis quelques essais que je fis sur la petite quantité que j'avois recue de Mr. Klaproth.*[32]

Vauquelin considered it highly unlikely that such a well-reputed chemist as Klaproth should take results from someone else. Vauquelin had certain doubts about Klaproth's analyses, since he had found a considerable amount of iron oxide in samples from him. In Paris the rumour spread that Hisinger and Berzelius had sent the mineral to Klaproth. This statement is given in Vauquelin's article in *Annales de Chimie* and in a translated form in the Philosophical Magazine.

> Some time after, Messrs. Berzelius and Hisenger, having been informed, by their correspondents at Paris, of M. Klaproth's labour, wrote to me to claim a priority, stating that they had sent to M. Klaproth the specimens of that mineral which he had used for his experiments, and that at the same time they had announced to him that they had found a new metal in it.[33]

The rumour also reached Klaproth's ear and he dispatched a critical letter to the unsuspecting Berzelius and asked for an explanation *"ob Sie dieses wirklich an Vauquelin geschrieben haben, oder nicht?"*[34] An answer to Klaproth to this question has not been found, but there is an extant draft from Hisinger's hand to Gehlen. Simultaneously a letter was sent with the same contents to Vauquelin to deny the allegation.[35]

As the publisher, Gehlen became the key person in the disputes about the priority. An exact point in time has not been established when he received each paper to be published in his chemical journal, but Gehlen gives Berzelius and Hisinger the honour of having discovered a new metal.

Es schien mir so in die Augen fallend zu seyn, dass die mit
Klaproth zugleich auf die Entdeckung Anspruch haben, dass ich,
indem ich am Ende Ihrer Abhandlung auf die Klaproth's zurück
wiess, bloss bemerkte es müsse überraschend seyn, denselben
Gegenstand von mehrern Seiten zugleich ins Auge gefasst zu
schen, und darauf aufmerksam mechte, dass Ihre (...) nung über
die metallische Natur der entdeckten Substanz die richtigere seyn
müsse, wenn sich die von Ihnen bemerkten
Oxydationserscheinungen bestätigten.
Klaproth will die Metallität der Cerits nicht annehmen; er spricht
immer von einem Mittelzustand zwischen Metall und Erde,
worunter ich mir nichts bestimtes denken kann.[36]

At the end of November 1804 Vauquelin received samples of Bastnäs tungsten to convince himself of the matter. With help from the chemical technician M Tassaert and mineralogist S Bergman the sample was analysed. The earlier results by Hisinger and Berzelius were fully confirmed.[37] Vauquelin thus seems to have had an intermediary role between Europe's foremost analyst (Klaproth) and the crownprince (Berzelius) in the field of chemistry. Klaproth eventually recognized the Swede's results of the analyses but kept the name of cererium which he regarded as more correct. Berzelius tried to renew the friendship with Klaproth, but in vain. It can look like an irony of history that, in 1817, Berzelius was offered the position of professor of chemistry in Berlin to succeed Klaproth.

2.3.2. THE QUANTITATIVE ANALYSES

In the first publication on the discovery of the new metallic oxide, Hisinger and Berzelius point out that their analysis first of all should be regarded as more qualitative than quantitative. They describe the properties of the different cerium compounds. The analyses of the cerite can be summarized as follows :

	Hisinger (1804)	Hisinger (1810)	Klaproth (1807)	Vanquelin (1805)
Ceroxide	50	68.59	54.50	63
Silica	23	18.00	34.50	17.5
Iron oxide	22	2.00	3.50	2
Lime	5.5	1.25	1.25	3-4
Volatile parts	?	9.60	5	12
Density	4.714-4.935	4.7-4.935	4.660	4.530

Gahn had measured the density of two different cerite specimens and found 4.781 and 4.981 g/cm^3 respectively. Cronstedt had come to 4.988 g/cm^3 but had considered his specimen to be contaminated. The mineralogical formula of cerite from Bastnäs was specified as *ceS + Aq* or Ce^3Si^2 + *6 Aq in* 1820 while in 1843 it was specified as *CeS + Aq or Ce³Si + 3 Ḣ* (a stroke represented two atoms ("double atoms"), oxygen was represented by a dot). Berzelius also introduced the mineralogical symbol *ce* for the cerous oxide and *Ce* for the ceric oxide. However, mistakes were frequent due to confusion between the mineralogical (in italics) and chemical symbol systems. Therefore the mineralogical symbols soon disappeared.[38]

Hisinger determined the composition of the oxides through forming the chlorides. The gravimetric determination was then made through precipitating the chloride ions with silver nitrate and the cerium ions with carbonate. The calculations were based on the fact that hydrochloric acid could neutralize a base containing 29.45 % oxygen. From this Hisinger got the composition of the cerous oxide to 85.08 % cerium and 14.82 % oxygen.[39] The calculations were furthermore based on Berzelius's theoretical statement that the ceric oxide should contain 50 % or 100 % more oxygen than the cerous oxide. This gave the formula for the cerous oxide CeO and the ceric oxide Ce^2O^3. The cerous oxide *(Oxidum cerosum,* C̈e) in modern terminology corresponds to the trivalent oxide Ce_2O_3. *Oxidum cericum* (often designated as the ceroxide) gives then the quadrivalent oxide CeO_2. In the year 1814 Berzelius determined the atomic weight of cerium to 1148.8 (O=100). The cerous oxide, designated Ce gets the formula weight 1348.8 which gives the proportion of cerium 85.2 % and the proportion of oxygen 14.8%. The agreement with experimental data is amazingly good. The ceric oxide (C̈e) contains theoretically 79.3 % cerium and 20.7 % oxygen. With the present value for the atomic weight of cerium being 140.12, we get for Ce_2O_3 cerium 85.4 % and oxygen 14.6 % by weight. The corresponding value for CeO would be 89.8 % cerium and 10.2 % oxygen. The quadrivalent oxide (CeO_2) has a proportion of cerium of 81.4 % and a proportion of oxygen of 18.6 %. Hisinger got in his experiments 79.3 % cerium and 20.7 % oxygen. These results give a plausible explanation to the fact that cerium was regarded to be a di- and trivalent metal. Uncertainty about the oxidation state of cerium persisted well into the 1870s.

It would be of interest to study the changes of the atomic weight of cerium during the years: *1814*: 1148.8; *1818*: 1149.4; *1827*: 574.77; *1836*: 574.70 and *1870*: 575 (O=100). Many atomic weights were changed with an even multiple in the mid-1820s.

Several unsuccessful experiments were then made to reduce the cerium compounds with carbon, potassium or through galvanic current. C G Mosander obtained a highly impure brown powder after passing potassium vapour over anhydrous cerium(II) chloride. F Wöhler also obtained an impure sample. W F Hillebrand and T H Norton succeeded in 1875 in producing a relatively pure metal through electrolysis of cerium(III)chloride[40] and later (1911) A Hirch did the same thing.[41]

Cerium has the relative atomic mass 140.12, atomic number 58, melting point 799°C, boiling point 3426°C, density 6.67 - 8.23 g/cm³ (due to allotrophy), and oxidation number 3 or 4. Cerium is the most common of the rare earth metals and can be found in the minerals monazite, bastnäsite, cerite and samarskite.

2.4. Cerine

Besides the common cerite, Berzelius and Hisinger had found in the mine of Bastnäs something which, on the surface, resembled amphibole but which in a preliminary analysis proved to contain ceroxide. To call attention to that fact, Hisinger named this mineral *cerin*. Hisinger's quantitative analysis had been so carefully described that it can illustrate the interpretation of such mineralogical analyses. The table of the results shows the composition of cerine:

Silica	0.314		
	1.270	1.584	30.17
Alumina	0.005		
	0.590	0.595	11.31
Lime		0.479	9.12
Ceroxide		1.480	28.19
Iron oxide	0.015		
	1.073	1.088	20.72
Copper oxide		0.046	0.87
Volatile parts		0.021	0.40
		5.293 g	100.78

The copper oxide was considered to be an impurity in the mineral. Since the sum of the included components is somewhat above 100 % the metallic oxides have their lowest oxidation numbers. Hisinger is not certain about the mineralogical formula and leaves the question open if it is a mixture of $CS + 2AS$ with $ceS + feS$.[42]

The analysis gave 11.31 parts alumina and 9.12 parts lime, whose oxygen is 5.28 and 2.56 parts respectively i.e. a ratio of 2 to 1. Cerine can thus be regarded as a double silicate of alumina, and lime $CS + 2AS$. The cerium(III) and the iron(II)oxides are similar. Berzelius was of the opinion that these latter substances were mechanically mixed into the mineral. He used the same argument for gadolinite and orthite.

The English chemist T Thomson reported that he had analysed a mineral with the same external characteristics and the same quantitative composition, which he called allanite. It might be worth mentioning that he believed that he had found a new element in the allanite, which he called junonium. Berzelius assumed a very sceptical attitude towards this novelty.[43]

2.5. Other Cerium-Containing Minerals

Jacob Berzelius often spent the summer months in the 1810s with Johan Gottlieb Gahn in Falun. The eagerness to find new minerals was great but there seemed to be too little time. Seldom Berzelius expresses his joy so openly and so unreservedly as in his descriptions of these visits. In a letter to Gustaf Brandel, a friend from his youth, in the late summer of 1814 we can read:

> Our chemical mineralogical research has taken an especially interesting turn and I am certain that if I was not obliged to be home till the beginning of October then I would not be home for Christmas. We have not got time to finish what we have started. I will return with so many minerals that there would be enough to pave whole Ladugårdslandstorget.[44]

In a letter of thanks to his master and host, Berzelius describes a happy time when he could entirely devote himself to the chemical-mineralogical studies:

> A thousand thanks for all the happy days in Fahlun, for all the minerals, for all the pleasant and instructive things we did together, for all the friendship and for all the kindness I enjoyed with you, sir, and your charming family during my almost 3-month-long stay in Fahlun.
> In my dreams I am still collecting rocks and boiling on the sandbath in the drawing-room, asking the maid if it is the 2nd or 3rd call for dinner and so on. - Alas, why could we not all move here together? Reality always surpasses dreams.[45]

The original samples of lanthanoxide and ceroxide. Berzelius Museum.
The Royal Swedish Academy of Sciences, Stockholm

Berzelius was in Falun again in the summer of 1816 and had some blasting done at Finbo, Broddbo and Kårarvet together with J G Gahn, C A Wallman, H P Eggertz, C G Gmelin and N Nordenskiöld. Several cerium-containing minerals were collected such as gadolinite, yttrotantalum, yttrocerite, difluate of ceroxide and yttria, basic fluate of ceroxide, neutral fluate of ceroxide and orthite.

When analysing gadolinites, Berzelius found that they contained cerium(III)oxide. This had entirely escaped earlier chemists' observations, but that is probably due to the resemblance between cerium(III)oxide and yttria. Berzelius had observed a brown residue after treatment with strong heat of what he thought was "the yttria". He suspected that the residue had emanated from a mistake in the precipitation of the iron with ammonium succinate. He produced a number of salts and could compare them with the corresponding salts of yttria. There were differences and he believed that he had proved the presence of cerium(III)oxide in gadolinites.

2.6. Epilogue

As a young scientist Berzelius already showed an impressive originality and an unusual clearness in argumentation and linguistic brillance. He proceeded methodically to find possible sources of errors in his analyses. He had a good pen and his stylistic ability to describe his chemical experiments in a short, simple and correct way was amazing. The difference in exactness and skill is great even in comparison with the more experienced chemists Johan Afzelius and Anders Gustaf Ekeberg. Ekeberg was often sceptical of Berzelius's chemical analyses.

The young Berzelius learnt a great deal from the episodes connected with the discovery of cerium. It appears clearly from all the letters, that the young scientist was enormously mature in his scientific argumentation and he had a great knowledge of the reactions of the different compounds. All the attacks on him for different reasons made him lose confidence in, and respect for, the established world and his self-confidence made a big step forward. Why should he be afraid of men like Klaproth, Vauquelin, Ekeberg and other scientific celebrities? Here we can already notice a new trend with clearly defined questions and answers, based on stricly empirical data. A great many experiments had to be repeated until the analyst had gained the skill which was necessary for judging the results. Here a great deal of patience was needed, a quality that Berzelius obtained as time went by. Here are some words of comfort when Sir Trolle-Wachtmeister complains about his being obliged to repeat an analysis for the eighth time in November 1819:

> About the analysis that you, Your Excellency, have started all over for the eighth time. I can only say that I like the repetition, this cannot happen too often as the wife said when her husband went to bed with her and to comfort you, Your Excellency, with regards to the possibility of a ninth time, may I mention that the late Mr Ekeberg wrote in his notes of his analyses that the

> gadolinite was analysed no less than 23 separate times and in all
> this a proportion of 16% ceroxide escaped his attention. - For if
> they do these things in a dry tree, then the green one must be
> content with what is done in it.
> I do not know if it helps Your Excellency if I confess that this
> very evening I am going to start my fourth analysis of potassium
> ferricyanide? The fifth is already coming towards me with open
> arms.[46]

There was no room for German natural philosophical speculations or half-hearted experimental results. The demand for accuracy and precision increased fast, which is well illustrated by the accuracy of the analyses of the cerium-containing minerals. As a logical consequence of this trend they could, from these minerals, isolate the rare earth metals which were very similar to cerium. Berzelius had made himself a name in the scientific world! He had come to stay - for a long time.

2.7. Acknowledgement

I would like to acknowledge Mr Kennet Flennmark, Lund for his translation of my Swedish manuscript and for rewarding linguistic discussions.

2.8. Abbreviations

KVAH = The Transactions of the Royal Academy of Sciences, Stockholm
KB = The Royal Library, Stockholm
KTH = The Royal Institute of Technology, Stockholm
KVA = The Royal Academy of Sciences, Stockholm
UUB = University Library, Uppsala
Crells Annalen = Chemische Annalen für Freunde und Naturlehre, Arzneygelahrtheit, Haushaltungskunst und Manufakturen
Brev = Jac. Berzelius Letters (brev) published by The Royal Academy of Sciences through H.G. Söderbaum, I-VI, Stockholm (1912-32) + supplement.
Letters to and from Berzelius are kept at The Royal Academy of Sciences in Stockholm if not otherwise specified.

2.9. References

1. A.F. Cronstedt, Rön och försök giorde med trenne järnmalms arter, KVAH   (1751), 226-32.

2. Anonym (A.F. Cronstedt), Försök till en mineralogie eller mineralrikets upställning, second edition, Stockholm (1781), 217-19.

3. C W Scheele, Tungstens bestands-delar, KVAH (1781), 89-95.

4. T Bergman, Tilläggning om tungsten, KVAH (1781), 95-8.

5. M E Weeks, The scientific contributions of the de Elhuyar brothers, J. Chem. Soc. (1934), 11, 413-9.

6. M H Klaproth, Untersuchung des angeblichen Tungsteins und des Wolframs aus Cornwall, Crell´s Annalen, II (1786), 502-7.

34   **Discovery of Cerium**

7.   C W Scheele to W Hising January 18, 1782 (A E Nordenskiöld, Carl Wilhelm Scheele - efterlämnade bref och anteckningar, Stockholm (1892), 403-5). See also C W Scheele to T Bergman May 31, 1782.

8.   T Bergman, Mineralogiska anmärkningar, KVAH (1784), 109-22.

9.   S Rinman, Bergwerks lexicon, II (1789), 248-9.

10.   B R Geijer, Crell's Annalen, I (1788), 229.

11.   J Gadolin, Undersökning af en svart tung stenart ifrån Ytterby stenbrott i Roslagen, KVAH, XV (1794), 137-55.

12.   A G Ekeberg, Ytterliggare undersökningar af den svarta stenarten från Ytterby och den däri fundna egna jord, KVAH, XVIII (1797), 156-64.

13.   A G Ekeberg, Uplysning om ytterjordens egenskaper, i synnerhet i jämförelse med berylljorden om de fossilier, hvari förstnämnde jord innehålles, samt om en ny upptäckt kropp af metallisk natur, KVAH, XXIII (1802), 68-83.

14.   J J Berzelius, Anteckningsbok från år 1807. KVA

15.   P J Hjelm (1746-1813), Torbern Bergman´s student, became Guard of the Mint in 1792 and head of the chemical laboratory at Bergskollegium, isolated molybdenum in 1781. W Hisinger to J G Gahn January 16, 1804. KTH

16.   J G Gahns efterlämnade skrifter. KTH (Gahn E13.2)

17.   W Hisinger to J G Gahn January 30, 1804.KTH

18.   W Hisinger to J G Gahn February 14, 1804. KTH

19.   J J Berzelius to J G Gahn February 15,1804 (Brev, supplement 2, 146-8).

20.   J G Gahn to J J Berzelius February 19, 1804 (Brev, IV:ii, 15-7).

21.   J G Gahns efterlämnade skrifter. KTH (Gahn E13.2)

22.   J J Berzelius to J G Gahn March 8, 1804 (Brev, supplement 2, 148-50).

23.   J J Berzelius to J G Gahn March 8, 1804 (Brev, supplement 2, 150).

24.   J J Berzelius to J Afzelius March 7, 1804 (wrongly dated 1809). UUB

25.   W Hisinger to J G Gahn April 26, 1804. KTH

26.   J J Berzelius to J Afzelius March 17, 1804. UUB

27.   M H Klaproth, Chemische Untersuchung des Ochroits, Neues allgemeines Journal der Chemie (1804), 303-16. The paper was also published separately and in the same year and it is among Gahn's extant manuscripts (Gahn E13.2, KTH). The paper was also published in other journals and books:
Analyse chimique de l'Ochroite, Mémoires de l'Académie Royale des Sciences et Belle-Lettres (1804), 155-64.
Chemical Examination of the Ochroits, a Mineral not hitherto well known, containing a new Earth, Journal of natural philosophy, VIII (1805), 207.

M H Klaproth, Beiträge zur chemischen Kenntnis der Mineralkörper, IV 1807), 140-52 (Chemische Untersuchung des Cererits).
See further G E Dann's Biography: Martin Heinrich Klaproth, Akademie Verlag, Berlin, 1958.

28.   W Hisinger to J G Gahn April 26, 1804. KTH

29.   The developed technique for analysis of inorganic elements including minerals has been well described in, inter alia, J J Berzelius, Lärbok i kemien, 2:a delen, Stockholm (1812), 4, 56-507; J Trofast, Excellensen och Berzelius, Atlantis, Stockholm (1988), 98-127; A I Samchuck och A T Pilipenko, Analytical Chemistry of Minerals, VNU Science Press (1987).

30.   Berzelius's and Hisinger's analyses of Bastnäs tungsten (cerite) was published in various publications:
Cerium, ein neues Metall aus einer schwedischen Steinart, Bastnäs Tungstein genannt, Neues allgemeines Journal der Chemie, Bd 2 (1804), 397-418.
Cérium, nouveau métal trouvé dans une substance minérale de Bastnas en Suède, appelée Tungstein, annales de chimie (1804), T. 50, An XII, 245-71.
Account of cerium, a new metal found in a mineral substance from Bastnas, in Sweden, Journal of natural philosophy, IX (1804), 290-300; X (1805), 10-12.
On cerium, a new metal found in a mineral substance of Bastnas in Sweden, called tungsten, the Philosophical Magazine, XX (1805), 154-8.
Undersökning af cerium, en ny metall, funnen i Bastnäs tungsten. Afhandlingar i fysik, kemi och mineralogi, I (1806), 58-84.

31.   L N Vauquelin, sur l'ocroite de M. Klaproth, lus à l'institut le 12 germinal an 12, Annales de Chemie, (1804), 140-3.

32.   L N Vauquelin to W Hisinger June 1, 1804. KVA

33.   L N Vauquelin, sur un minéral appelé autrefois faux Tungstène, aujourd'hui Cérite, et dans lequel on a trouvé un métal nouveau, Annales de Chemie, LIV (1805), 28-65.
Account of Experiments made on a Mineral called cerite, and on the particular Substance which it contains, and which has been considered as a new Metal, The Philosophical Magazine, XXII (1805), 193-200.

34.   M H Klaproth to J J Berzelius July 10, 1805. KVA

35.   The drafts of the letters to Gehlen and Vauquelin are dated Skinnskatteberg September 7, 1805. KVA

36.   A F Gehlen to W Hisinger May 22, 1804.KVA

37.   L N Vauquelin, Experiments on a Mineral formerly called false Tungsten, now Cerite, in which a new Metal has been found, Journal of Natural Philosophy (1805), XII, 105-11. The results in which the experiments to get pure metallic cerium are presented, can also be found in Annales de Chimie (1805), LIV, 28-65.

38.   W Hisinger, Handbok för mineraloger under resor i Sverige, Stockholm (1843), 36-7.

39.   W Hisinger, Undersökning af svenska mineralier: ceritens analys, Afhandlingar i fysik, kemi och mineralogi (1810), III, 283-9.
W Hisinger, Försök att bestämma syrehalten i ceroxidul och ceroxid (1813), KVAH, 216-218.

40.   W F Hillebrand och T H Norton, Pogg. Ann. 155 (1875), 631; 156 (1875), 466.

41.   A Hirch, The preparation and properties of metallic cerium, Met. Chem. Eng. 9 (1911), 540-4.

42.   W Hisinger, Analys av cerin, Afhandlingar i fysik, kemi och mineralogi, (1815), IV, 327-33.

43.   J J Berzelius, Bilaga till Herr Hisingers analys af cerin, KVAH (1811), 215-8.

44.   Ladugårdslandstorget = Östermalmstorg i.e. a large square in Stockholm. J J Berzelius to Gustaf Brandel September 8, 1814. KB

45.   J J Berzelius to J G Gahn September 26, 1814 (Bz Brev IV:ii, 100).

46.   J J Berzelius to H G Trolle-Wachtmeister November 30, 1819. Trolle-Ljungby. A part of the letter is a paraphrase of St Luk. 23:31.

CHAPTER 3

CARL GUSTAF MOSANDER AND HIS RESEARCH ON RARE EARTHS

LEVI TANSJÖ
Chemical Center, Lund University,
P.O. Box 124, S-221 00 Lund, Sweden

3.1. Introduction

Carl Gustaf Mosander's discoveries of lanthanum and didym in ceria and of erbium and terbium in yttria were communicated in a paper read at the 13th meeting of the British Association for the Advancement of Science held at Cork, Ireland, August 17-23, 1843. The title of the paper was *On the new metals, Lanthanium and Didymium, which are associated with Cerium; and on Erbium and Terbium, new metals associated with Yttria* (1). It was read by Mosander's brother-in-law, the Irish dragoon-major North Ludlow Beamish, who lived in Cork and was President of the Cork Scientific and Literary Society. He had also translated the paper into English. The paper attracted great attention. It was printed the same year in the October issue of the Philosophical Magazine and translations of it also appeared before long in German and French journals (1). The main part of the paper had been read by Mosander in Swedish under the title *Något om Cer och Lanthan* (Something on Cerium and Lanthanum) at the 3rd meeting of *Skandinaviska Naturforskare-Sällskapet* (Scandinavian Association) held in Stockholm, July 13-19, 1842 (2).

Mosander's discoveries induced many chemists to study minerals containing cerium and/or yttrium, but progress was very slow until chemists in the late 1870s learned how to handle the new sharp-edged weapon spectroscopy and improve their separation techniques. After that the progress was rapid. In 1907 the oxides of not less than 14 of the 16 elements between barium and tantalum had been discovered and isolated, in spite of their great similarity and the great rarity of some of them. This research, carried out without much guidance of Mendelejev's periodic table, was characterized by George de Hevesy in 1926 as "one of the most brilliant accomplishments that experimental chemistry has ever produced" (3). And he could have added that it had been brilliantly applied by Carl Auer von Welsbach in his gas mantle and pyrophoric alloys (See chapter 7).

Mosander was a pioneer in this glorious field of research!

C. H. Evans (ed.), Episodes from the History of the Rare Earth Elements, 37–54.
© 1996 *Kluwer Academic Publishers. Printed in the Netherlands.*

3.2. Biography

Carl Gustaf Mosander (10 September 1797-15 October 1858) was born at Kalmar on the Swedish east coast 400 kilometers south of Stockholm. His father Isac Mosander was a sea-captain, his mother Christina Maria was born Törnqvist. In 1809 he moved with his mother to Stockholm where, at the age of fifteen, he became an apprentice at the pharmacy Ugglan (The Owl). He passed a pharmaceutical examination and seemed bent on a career as druggist. His interest, however, turned to medicine and in 1820 he began medical studies at the Karolinska Institute, where Berzelius awoke his interest in chemistry. Already as a student he worked now and then in Berzelius's laboratory at the Academy of Sciences, where he met Friedrich Wöhler. They became close friends. Mosander passed his medical examination ("*kirurgie magister*") in 1825. He was then already appointed as teacher in chemistry under Berzelius at the Karolinska Institute. Berzelius appointed him in 1828 custodian of the Academy's mineral collection and he was given a laboratory of his own and a flat in the Academy's newly acquired building on Drottninggatan.

In 1832 Mosander succeeded Berzelius as professor of chemistry and pharmacology at the Karolinska Institute. Berzelius's retirement that year at the age of 53 was partly meant as his wedding present to Mosander. Such a generosity was then - and is still - not very common in Sweden.

From 1845 Mosander was professor of chemistry not only at the Karolinska Institute but also at the Pharmaceutical Institute, as well as being inspector of this Institute which meant yearly inspection visits to pharmacies all over Sweden. He spent much of his time organizing the mineral collections at the Academy of Sciences, and at the State Museum of Natural History in Stockholm. From 1825 he was also owner and manager of a spa-establishment in central Stockholm for artificial Karlsbader waters. These and many other activities did not leave much time for research. But he found time to flirt with Hulda Philippina Forsström, famous all over Stockholm for her beauty. They married on the 20th of December 1832. By January 1833, Mosander was still, as Berzelius expressed it, "lying pale and yawning all day long, anchored at the bosom of his beautiful bride" (4, January 3, 1833). They had four children, two pairs of twins. In 1833 two boys were born and in 1836 twins again: a boy and a girl (*Note 1*).

Mosander died in 1858 at the age of 61 in his summer house on the island of Lovön near Drottningholm castle. Since his forties he had suffered from cataracts and was almost blind during his last years. His beautiful wife died in 1888.

3.3. Mosander's Chemical Research

As mentioned before Mosander had many irons in the fire and published, in addition to his famous papers (1) and (2), only five papers in pure chemistry, running to a total of 62 pages.

I.   *Undersökning af så kallad jernsinter*, 1826 (5).
II.   *Undersökning af en Serpentin-art från Gullsjö*, 1826 (6).
III. *Något om Cerium*, 1827 (7).
IV.   *Undersökning af några arter Titan-jern*, 1830 (8).
V.   *Undersökning af några cyan-dubbelsalter*, 1834 (9).

In his first work (I) Mosander showed that so-called iron-sinter, the oxide film formed on iron when it is made red-hot in air, is not - as the French chemist Pierre Berthier had maintained (10) - a well-defined mixture of iron oxidule, FeO, and iron oxide, Fe_2O_3, in the mole ratio 2:1. Mosander found that it consists of two layers, the inner one much richer in oxidule than Berthier had found and the outer one of variable composition. In September 1824 (Paper II) Mosander analysed a serpentine species from Gullsjö in Northern Sweden and in March 1825 he did it for the eighth time "to learn not to work *geschwind* and *schlecht*" as Berzelius wrote to Wöhler, then in Berlin (11; March 15, 1825). In connection with this work Mosander found that when soda, Na_2CO_3, is used to precipitate talc earth as carbonate, a double-salt of soda and magnesium carbonate is formed, which contaminates the magnesium carbonate. By that he demonstrated that up to then all analyses of minerals containing magnesium had resulted in too high a percentage of magnesium.

(Paper III is accounted for in the next section).

In 1829 (Paper IV) Mosander found titanium-iron minerals to be mixtures in various proportions of iron oxide and a compound of iron oxidule and titanic acid in which 2/3rd of the oxygen is combined with the titanium. Mosander noticed that ilmenite and other titanium-iron minerals are isomorphous with iron oxide, which was easily explicable only if the titanic acid had the composition TiO_2, *i.e.*, the composition that Berzelius and others already had presumed without experimental support. Since titanic acid and stannic acid were known to be isomorphous Mosander through this work also gave support to the assumed formula SnO_2 for stannic acid (*Note 2*).

In his 28-page Paper V in 1833 - his first one as professor - Mosander accounted for his extensive study of the composition of the precipitates formed when 'yellow prussiate of potash', often called 'yellow Blutlauge', is added to various salt solutions. Yellow Blutlauge which, in the light of Alfred Werner's coordination theory, we call potassium hexacyanoferrate(II), $K_4[Fe(CN)_6]$, was considered to be a double-salt, $4KCN \cdot Fe(CN)_2$, by Berzelius and consequently by Mosander (*Note 3*). Mosander found that whenever this compound is used to separate other metals as precipitates from solutions, the potassium in the yellow Blutlauge is always replaced by another metal but *never completely*, *i.e.*, the separated precipitates are always 'triple-salts'. We know that the composition of the precipitate depends on the conditions of the experiment but a good formulation of, for instance, Prussian blue is $KFe[Fe(CN)_6]$. Some of the potassium in the yellow Blutlauge is always left, just as Mosander taught!

3.4. Mosander's Research on Rare Earths

3.4.1. BACKGROUND

The terms rare earth metals and alkaline earth metals - still in use in chemistry - have their origin in the word 'earth' without a plural, used by alchemists, iatrochemists and early phlogistonists as a general designation for fireproof, insoluble residues formed after combustion or other drastic chemical treatments of substances. Only after 1750 did chemists clearly distinguish *silica, lime, alumina* and *talc* (after its carbonate also called *magnesia*) as different earths, *i.e.*, as chemical individuals. To them Scheele added *baryta* which, in 1779, he definitely distinguished from lime. Lavoisier included these five earths in his list of 33 elements in *Traité élémentaire de chimie* in 1789 but remarked that they might be oxides of metals with a greater affinity to oxygen than carbon (*Note 4*). This induced many chemists in the early 19th century to try to reduce known earths and to search for new ones.

Six new earths were discovered in 1789-1828:

1789   zirconia (Klaproth)
1790   strontia (Adair Crawford)
1794   yttria (Gadolin)
1798   beryllia (Vauquelin, who called it glucina)
1803   ceria (Klaproth and independently Berzelius and Hisinger)
1829   thoria (Berzelius) (*Note 5*)

Ceria made the borderline between earths and ordinary metal oxides vague. All the earths known before 1800 were non-oxidizable, *i.e.*, they seemed - if they were oxides of unknown metals - to be the highest oxides of the metals in question. They differed, however, clearly from the highest oxides of manganese, chromium and some other metals since they were not acidic and seemed to be non-reducible. That made them to a well-defined group of substances. The new compound isolated from Bastnäs tungsten in 1803, however, was either oxidizable (if it was isolated as Ce_2O_3) or reducible (if it was isolated as CeO_2). The young Berzelius seems immediately to have realized that this could not be lodged in any reasonable definition of the concept earth and the new compound was therefore called a metal oxide and the new metal was named cerium. Its two oxides were called cerium (or cer) oxidule (in English protoxide) and cerium (or cer) oxide. Cerium and its compounds were already in the 1st edition of Berzelius's text-book *Lärbok i Kemien* (1808), treated not among the earths but among other metals (between manganese and uranium). The word *ceria* was in fact never used by Berzelius and his students, but many chemists continued to use it and to consider ceria as an earth just as Klaproth had done in 1803.

Since silica reacts with alkalis to give glass, Berzelius preferred to call it silicic acid and expelled also this substance from the group of earths. In the German translation of the 2nd edition of his text-book (1825) it was called *Kieselsäure* and not treated among the earths but among inorganic acids (as the last one). After that only nonvolatile, infusible substances which, like the alkalis and oxides of metals neutralise acids to give salts, were considered to be earths by Berzelius and his students. These were already in the 1st edition of Berzelius's text-book classified as *alkaline earths* and *true earths*.

The alkaline earths were marked by their tendency to give water solutions similar to those of the alkalis, *i.e.*, caustic soda and potash and the volatile alkali ammonia. *Lime, baryta, strontia* and - in spite of its slight solubility in water - *magnesia* were classed as alkaline earths.

To the true earths, which were thought to be quite insoluble in water, these were counted: *alumina, zirconia, yttria, beryllia, ceria* and, after its discovery in 1829, *thoria* (but not ceria in Berzelius's school).

In 1808 Davy reduced the alkaline earths by electrolysis to the alkaline earth metals calcium, barium, strontium and magnesium (*Note 6*). After that most chemists felt quite sure that the true earths were oxides of metals too, but all attempts in the 1810s to reduce them were in vain. In the 1820s, however, all the expected metals were obtained - but impure and in small quantitites - by treatment of their fluorides or chlorides with metallic potassium or its vapour:

1824	zirconium (Berzelius)
1825	aluminium (Oersted; better by Wöhler in 1827)
1826	cerium (Mosander)
1828	beryllium (Wöhler and Antoine Bussy independently)
1828	"yttrium" (Wöhler)
1829	thorium (Berzelius)

The volatile chloride, however, from which Wöhler in 1828 prepared what he thought to be metallic yttrium (12), was most certainly beryllium chloride. Heinrich Rose found in 1843 that the chloride prepared from beryllia-free yttria was not volatile. From yttrium chloride or the corresponding fluoride he obtained metallic yttrium by reduction with metallic sodium (13).

3.4.2. SOMETHING ON CERIUM

When Mosander had finished his first two works (I and II) in the summer of 1825, Berzelius asked him to find out whether 'sulphur cerium', *i.e.*, a sulphide of cerium, is formed when cerium oxidule or oxide are strongly heated in an atmosphere of carbon disulphide and within a few weeks he made what Berzelius in a letter to Wöhler called "most remarkable things on cerium" (11; September 25, 1825). Mosander obtained the 'sulphur cerium', corresponding to the oxidule (Ce_2S_3) both from the oxidule in vapour of carbon disulphide and from the oxide when this was heated with 'hepar' (potassium polysulphide). And this new compound gave with chlorine the *anhydrous* chloride $CeCl_3$, which - in contrast to the hydrates of cerium halides formed on crystallization from water solutions - was easily reduced by potassium vapour to give, for the first time, cerium in metallic form. Mosander found it to be a most reactive metal. Even at $0°C$ it decomposed water, which made it difficult for Mosander to remove the potassium chloride formed together with the metal in the reduction process. When he removed it with water he instead obtained a mixture of cerium and its oxidule since then some of the metal reacted with the water. The best results were obtained when he used 75 % ethanol instead of water to dissolve the potassium chloride. In the same work Mosander also prepared a selenide, a phosphide and a carbide of cerium.

Mosander mentioned in his lecture at the Stockholm meeting in 1842 that he became convinced that the cerium oxidule he had used was not a pure oxide but a mixture of metal oxides. He had already arrived at this conclusion in connection with his work on cerium in 1825-26, published under the title *Något om Cerium* (Something on Cerium). He was, however, unable to separate the components of this mixture and gave it up when he had spent all his cerium oxidule in his efforts to do it.

During the next few years Mosander often mentioned to Berzelius that "all was not right with cerium" (4; February 5, 1839), but twelve years passed before he prepared a new sample of cerium oxidule and set about to work at the problem. A fuller account of the discovery of cerium is given in the preceeding chapter.

3.4.3. THE DISCOVERY OF LANTHANUM

In the autumn of 1838 Mosander was stuffing small blocks of cerite and cerine into pieces suitable for mineral collections, a process which generated a considerable amount of waste. From this one of his students prepared 3 *skålpund* (1.3 kg) of a double-salt of potassium and cerium sulphate, from which Mosander prepared cerium oxidule and a number of its salts. Studying these he became still more convinced than he had been in 1826 that they contained more than one metal. His efforts to separate them, however, were again in vain until he got the bright idea that the metal oxide which he supposed to be mixed with the 'true' cerium oxidule might be more basic than the highest oxide of 'true' cerium, *i.e.*, cerium oxide. That turned out to be the case and Mosander could now extract from cerium oxide the supposed unknown metal with chlorine water as chloride, and with very dilute nitric acid as nitrate. The oxide he obtained from these salts was neither oxidisable nor easily reducible to a lower oxide, *i.e.*, it differed completely from the oxides of 'true' cerium and was not identical with any known metal oxide.

Carl Gustaf Mosander (Courtesy Library, Karolinska Institute Stockholm)

A new element was discovered and it was high time for it! No new element had appeared after Gabriel Sefström's discovery of vanadium in 1830.

It appears from Mosander's laboratory diary "C.G. Mosanders Laborations-Bok", that he discovered the new oxide in November 1838 (*Note 7*). He started his first successful splitting of cerium oxide on November 20 and it seems as if five days later, on November 25, he realized that he had found a new oxide; but he did not blow his own trumpet about it. He had certainly succeeded in splitting cerium oxide into two fractions but he knew quite well that the separation was far from complete and the new oxide seemed to be contaminated with much more than a residue of cerium oxide. He did not even tell Berzelius about his discovery, in spite of the fact that they lived and worked in the same house and met daily. It might also have been a little delicate to inform Berzelius, since the discovery meant that Berzelius had been wrong about cerium for the past 35 years! At Christmas-time, however, Mosander was provoked to tell him.

Axel Erdmann - a young co-worker of Sefström in Falun - had sent 5 mg of an oxide to Berzelius and asked him to investigate it with the blow-pipe. Erdmann had isolated it from a new mineral found at Brevik, Norway, and he thought that it was the oxide of a new metal. When Berzelius told Mosander about this at Christmas-time and showed him Erdmann's oxide, Mosander became anxious that it was identical with his own new oxide and gave Berzelius an account of his discovery and one gram of the new oxide to study. Berzelius first hoped that Mosander was wrong , but already during the Christmas holidays he realized that Mosander really had found a new oxide in cerium oxide, overlooked by himself and Hisinger in 1803, and that it was quite different from Erdmann's oxide (14; January 3, 1839) (*Note 8*). Berzelius asked Mosander to name his 'foundling', and proposed himself the name *lanthan* for it from the Greek word for "hidden", since the metal is always hidden by cerium. Mosander accepted this name in the beginning of February 1839 (11; February 12, 1839). In his laboratory notes he still called it on February 3 "X-oxid" but three days later, on February 6, "lanthan oxid". Both Mosander (in his notes) and Berzelius saw it first as an earth. Berzelius emphasized its resemblance to magnesia and meant that it could be classed as an alkaline earth. Before long, however, both Berzelius and Mosander did not see lanthanum oxide as an earth but as an ordinary metal oxide.

When isolated from cerium oxide, lanthanum oxide was easily distinguished from the two cerium oxides, but when mixed with them it changed their properties very little. Berzelius and Mosander wrongly thought that the two cerium oxides had the formulae CeO and Ce_2O_3, which they wrote $\dot{Ce}$ and $\ddot{Ce}$. They thought lanthanum oxide to be $\dot{La}$, *i.e.*, LaO.

Mosander's discovery of lanthanum became quickly known through Berzelius's letters about it to colleagues abroad in February 1839. Excerpts from his letter to Jules Pelouze in Paris, dated February 22, appeared as notices in *Comptes rendus hebdomadaires des séances de l'Academie des sciences de Paris* and elsewhere (*Note 9*) and when the Royal Swedish Academy of Sciences on April 2, 1839, celebrated its centenary Berzelius reported the discovery of lanthanum in his "Årsberättelse" concerning the year 1838 (*Note 10*). Mosander was on that occasion awarded the so-called Lindbom-prize for his discovery. In February he had prepared lanthanum sulphide, reduced anhydrous lanthanum chloride with metallic potassium to lanthanum in metallic form and found its atomic weight to be lower than that of cerium. Berzelius reported this news in a letter, dated May 3, 1839, to J.C. Poggendorff, who published an excerpt of it in his *Annalen* (*Note 11*).

Wöhler waited eagerly for a full paper from his friend Mosander on lanthanum to be published in *Annalen der Chemie und Pharmacie*, which he edited together with Liebig (*Note 12*), but "Pater Moses" - Berzelius's and Wöhler's nickname for Mosander - was not ready for publishing anything about lanthanum. Instead he continued to study - far from daily, but occasionally when he had time for it - his new oxide and before long he found it to be a mixture of many oxides. It contained not only residues of cerium oxide, but also lime and oxides of iron, copper, tin, nickel, and something resembling uranium. When separated from these oxides, the lanthanum oxide was brick-red in colour but it became dirty white when heated to white heat in air. It lost its colour also when heated in hydrogen gas, without undergoing a perceptible loss of weight. These phenomena and the fact that the intensity of its brick-red colour shifted from experiment to experiment caused Mosander to presume that his lanthanum oxide and the reddish-brown cerium oxide, left after the extraction of lanthanum oxide from it with nitric acid, were accompanied by some unknown oxide. That turned out to be the case!

3.4.4. THE DISCOVERY OF DIDYMIUM

In January 1840 Mosander succeeded in splitting his then amethyst-coloured lanthanum sulphate into two fractions. When he heated a solution of this salt - already known to be more soluble in cold than in warm water - from 9 to 40 °C light amethyst-coloured crystals precipitated. When they were treated in the same way, colourless crystals were obtained after 10-15 operations. With alkali, and subsequent heating of the oxide-hydrate formed, they gave a *white oxide*, apparently the 'true' lanthanum oxide. Mosander called it in his note-book $\overset{\bullet}{L}a^a$. From the amethyst-coloured solution he obtained red crystals, apparently containing the sulphate of the unknown oxide. In his note-book Mosander called this oxide $\overset{\bullet}{L}a^r$, where r presumably = röd (red). The red sulphate crystals gave with alkali a bluish-violet oxide-hydrate which on heating lost water leaving an oxide,

> "dark brown on the surface, sometimes light brown in the fracture,
> of a resinous lustre, sometimes nearly black, with the lustre and
> appearance of dark orthite, at the same time particles are obtained of
> all the most dissimilar colours, so that they represent together a
> pattern map of all the most dissimilar kinds, which are obtained of
> the mineral orthite, from the light red brown to the nearly black" (2).

When heated to a white heat it assumed a dirty white colour approaching grey green. It turned out to be a weaker base than lanthanum oxide and gave amethyst-red salts.

Mosander could from now on prepare tolerably pure, almost white lanthanum oxide and light yellow cerium oxide and prove that the amethyst colour often attributed to the salts of these oxides, was caused by his La^r-oxide and also the brown colour which these oxides used to assume when heated to red heat in air. He found the "atomic weight" of his purified lanthanum oxide to be 680 (O = 100) corresponding to an atomic weight (O = 16) of lanthanum = 139.2 (modern value 138.9) (*Note 13*).

Mosander later named the unknown metal in the La^r-oxide *didymium* after the Greek word for twins, because it had been discovered, *i.e.*, born, in conjunction with lanthanum just as twins are born together. To its oxide he ascribed the formula Ḋi, *e.g.*, DiO (*Note 14*).

The amount of crystal water in the red sulphate from which Mosander had obtained the La^r-oxide convinced him that this salt in fact was a double-salt of didymium oxide and something else, presumably lanthanum oxide. He did not succeed in splitting it and consequently had to consider his La^r-oxide as a compound of didymium oxide and presumably lanthanum oxide. Neither did he succeed in splitting the double salts he obtained from La^r-oxide and acids other than sulphuric acid. Mosander saw, therefore, no reason to study seriously his La^r-oxide, or as he clearly expressed it at the Stockholm meeting:

> "When we find the oxide of a hitherto unknown body, nothing,
> generally speaking, is easier than the determination of the qualities
> of the body. That which, in the first place, gives any
> value to chemical investigation, is the certainty that the object is
> pure, that is to say, free from other substances".

Mosander did never get his didymium oxide tolerably pure.

Mosander met Berzelius almost daily and complained often in their talks about "the difficulties, which prevented him to obtain pure preparations" (11; May 13, 1842), but he did not inform Berzelius about his discovery of didymium until the beginning of 1841 and then under promise of secrecy, which Berzelius kept (11; August 2, 1842). In the spring of 1842 he also informed the **lecturer in mineralogy and metallurgy at the** University in Christiania (now Oslo), Theodor Scheerer.

Theodor Scheerer (1813-1875) was a German chemist, a pupil of Wöhler. Before he became professor at the Academy of Mines at Freiberg in Germany in 1848, he worked in Norway, from 1833 at the cobalt-mine at Modum and from 1841 as university lecturer. He was an eminent mineralogist and made a number of accurate analyses of minerals. In connection with an analysis of gadolinite he observed some peculiar colour changes when yellow yttria was strongly heated under various conditions. Scheerer first thought that the changes were caused by lanthanum oxide contaminating his yttria. When this explanation was found to be untenable he instead assumed the presence in his yttria of an unknown oxide, which he thought might be the cause of the colours of both cerium oxide and lanthanum oxide. This idea he put forward to Berzelius, who thought that the whole thing had something to do with Mosander's didymium oxide. Berzelius arranged a meeting in Stockholm between Mosander and Scheerer in the spring of 1842, at which they openly discussed their experiences and agreed that both of them should read a paper on their works at the meeting of the Skandinaviska Naturforskare-Sällskapet in July 1842 (11; August 2, 1842).

412 scientists and physicians from Denmark, Norway and Sweden and 24 from other countries participated in the meeting of the Skandinaviska Naturforskare-Sällskapet in Stockholm 13-19 July, 1842. Berzelius, who was chairman of the Association, opened the meeting with a brilliant speech in which he took the opportunity to exhort scientists to be very circumspect with hypotheses in science. He finished the speech with these words:

> "It is easy to travel through spaces on the Icarus wings of hypotheses; but the sun will sooner or later melt the wax of the wings. The Icarus tale was made as a warning to practicians of sciences too. Let us not forget that tale. May thorough inquiry constitute the principal feature of the works in common we today have begun" (15).

H. C. Oersted was elected chairman of the section for physics and chemistry in which Scheerer and Mosander read their papers on July 16, Scheerer first and then Mosander. Scheerer had divided his paper, read in Norwegian, into two parts: a) *Chemisk Undersögelse af Gadoliniten fra Hitteröen og af et andet Mineral fra samme Findested* (Study of the gadolinite from the island of Hitteröen and of another mineral from the same finding place) and b) *Nogle chemisk-analytiske Erfaringer, samlede ved Analysen af de naevnte Mineralier* (Some chemical-analytical experiences, collected at the analysis of the minerals mentioned) (16). Scheerer suggested cautiously in his paper that yttria might contain an unknown oxide but he did not develop the idea further. Mosander connected his paper *Något om Cer och Lanthan* (Something on cerium and lanthanum) to Scheerer's one with the introduction:

> "Although in consequence of the imperfect nature of the results which were obtained from my researches on cerium and lanthanium I had no intention of making any communication on the subject on the present occasion, yet after hearing the interesting lecture of Lector Sche[e]rer, it appeared to me that it might be useful to make known more generally some particulars which arose during my labours, and principally because this advantage may result, that other chemists, after becoming acquainted with what I am about to state, may possibly be spared the loss of valuable time which might otherwise have been fruitlessly expended".

Then he accounted for his discovery of lanthanum oxide and his studies of various salts of tolerably pure cerium oxides and lanthanum oxide. The last third of his lecture he devoted to his discovery of didymium oxide of which nobody in the audience except Berzelius and Scheerer had the slightest idea. Mosander, in contrast to Scheerer, did not think that didymium oxide was the unknown oxide contaminating yttria but finished his lecture with these words:

> "I must not omit to mention on this occasion, that amongst the many
> other bodies which in the course of these researches I was obliged to
> examine, yttria also presented itself, and I have found that this earth,
> free from foreign substances, is perfectly colourless, and gives
> perfectly colourless salts; that the amethyst colour which the salts
> generally present comes from didymium, I will not, however,
> maintain."

Mosander's discovery of didymium became quickly known among chemists after
Scheerer's report on it - with Mosander's permission - in Poggendorff's *Annalen* and
Wöhler's one in *Annalen der Chemie und Pharmacie* (*Note 15*).

3.4.5. THE NAME DIDYMIUM

The Stockholm meeting was a great success, but a trying experience for the aging
Berzelius. During the week before the meeting he had to entertain prominent guests, who
arrived early; during the meeting he was busy every day but felt yet obliged to stay until
late in the evenings in the club arranged for the participants in the Palace of the Crown-
Prince (now Ministry of Foreign Affairs); and during the week after the meeting he had to
entertain prominent guests who thought that one week in Stockholm was not enough. On
August 2, however, he had recovered from the hardships of the meeting and gave, in a
long letter to Wöhler, a full report of both its scientific and non-scientific parts. He
praised Mosander's "not unpretentious" lecture and his confidence while performing it.
Wöhler was glad that "Pater Moses" again had discovered a new element but he did not
like the name *didym* for it, by which he meant it sounded childish and ridiculous in
German. He asked Berzelius to persuade Mosander to propose a better name before
didym came into print, but Berzelius refused to execute that mission and answered:

> "...I have no share in that name and I am not willing and not able to
> ask Mosander to change it, after he has announced it publicly. You
> surely do not really know our friend Pater Moses. He takes
> suggestions from no one. The proposal to change a name given by
> him would be an offense which he would not easily pardon, and still
> he would not change it. He intentionally looked for a name
> beginning with D in order to get a symbol unlike those for other
> metals. It is certainly quite true, as you say, that the repetition of the
> same consonants, and of almost the same vowel sounds, is not
> euphonious; but one soon gets accustomed to it, and finds it
> endurable, and you must do the same. What does furthermore didym
> matter against acide imasitinasique, chloronaphthalase,
> naphthalidam *etc.*, with which we at present yet have to be satisfied
> ...." (11; September 6, 1842).

Wöhler did not agree that the excellent atomic symbol D compensated for the flaws of the
name didym and added in his next letter that Moses apparently had chosen a name

meaning twins "in remembrance of the lively pair of twins which he has made with his beautiful wife in the wedding-night" (11; September 16, 1842).

The name didym or didymium survived Wöhler but not for long. It disappeared when Auer von Welsbach in 1885 succeeded in splitting didymium oxide into two oxides. (See following chapter and chapter 7). Mosander and after him J. C. G. Marignac, P. T. Cleve, F. Nilson, and F. Lecoq de Boisbaudran had, under similar circumstances, kept the old name for one of the components in the mixture but Auer von Welsbach was more pretentious. He gave new names to the metals in both of them - *neodymium* and *praseodymium* - and is credited as discoverer of both of them. He had less success with the names *aldebaranium* and *cassiopeium* for the metals in the two oxides into which he split Marignac's ytterbia in 1907. The International Atomic Weight Committee preferred the names *ytterbium* and *lutecium* (from 1949 lutetium) for them (see chapter 5). In doing this the committee followed the tradition that Mosander and Berzelius founded in February 1839 when cerium oxide had been split: one of the metals they called lanthan but they kept the name cerium for the other.

3.4.6. THE DISCOVERY OF TERBIUM AND ERBIUM

Scheerer, and many with him, thought after Mosander's lecture at the Stockholm meeting that ordinary yttria contained didymium oxide. However, in the autumn of 1842 Mosander became quite sure that samples of yttria, isolated from the minerals gadolinite, cerite, cerine, and orthite, all were mixtures of the 'true' white yttria giving colourless salts, and an unknown, less basic, yellow oxide, giving amethyst coloured salt solutions. In February 1843 he intended to name the new metal in this oxide *odinium*, (11; February 16, 1843), obviously after the ancient Nordic god *Odin,* but subsequent experiments - his laboratory notes suggest in April - convinced him that there were in fact at least three oxides in yttria. When he fractionally precipitated basic salts with ammonia from a solution of ordinary yttria in nitric acid, precipitates were formed which on strong heating left three, distinguishable oxides, none of them being identical with a known one. Berzelius confirmed the discovery that yttria could be split into three oxides and reported it in his "Årsberättelse" for 1842, which in April 1843 had not yet been printed. Mosander found then that the three oxides could also be separated by fractional precipitation as oxalates.

For the most basic oxide, which was white and gave colourless salts, Mosander kept the old name *yttria*. The next one in basicity he named oxide of *terbium.* It was assumed by Mosander to be white when pure, but it furnished amethyst red salt solutions. In both his methods of separation Mosander found yttria in the final fractions and the oxide of terbium in the central ones. The third oxide, the least basic one he named oxide of *erbium.* On heating in air it became dark orange yellow, a colour which it lost, with a trifling loss of weight, when heated in hydrogen gas. That indicated the existence of two erbium oxides, and that the yellow component in ordinary yttria in fact was a compound of erbium oxide with many 'atoms' of erbium oxidule.

Mosander's names terbium and erbium on the two new metals were, like yttrium and yttria, derived from Ytterby, the name of the small village near Stockholm, where Carl Axel Arrhenius had in 1787 found the mineral later called gadolinite.

In May 1843 Mosander gave Berzelius a note and verbal accounts of his research on yttria and Berzelius confirmed many of his experiments. The section on yttria in the 5th edition of Berzelius's *Lehrbuch der Chemie* (1844) contains what Berzelius and Mosander knew about yttria when Berzelius on June 2, 1843 sent the manuscript for the section in question to Wöhler (11). Mosander's discoveries amazed Berzelius. In 1802 A.G. Ekeberg had found beryllia in Gadolin's 1794-yttria and Berzelius himself had, in 1814, found cerium in it. Mosander's discoveries meant that in 1794 Gadolin had in fact discovered a mixture of oxides of seven then unknown metals: beryllium, cerium, didymium, lanthanum, erbium, terbium, and yttrium. Berzelius was amazed and thought: "What a scoop it would have been if he [Gadolin] had been able to separate them" (4; June 13, 1843).

3.5. The Cork Paper

Mosander did not announce his discovery of erbium and terbium in a journal but in the paper that his brother-in-law N.L. Beamish read at the meeting of the British Association at Cork in August 1843. This paper was,however, without any effort on the part of Mosander, published in not less than five journals before it appeared in the meeting report. The main part of the paper was just a translation, made by Beamish, of the paper *Något om Cer och Lanthan*, that Mosander had read at the Scandinavian Association meeting in Stockholm the year before. Beamish had participated in that meeting and in the summer of 1843 he was again visiting Stockholm. Mosander then wrote a brief report on his discovery of erbium and terbium, which Beamish also translated. It was appended to the translation of Mosander's lecture at the Stockholm meeting as *Addendum, July 1843 - On Yttria, Terbium and Erbium*. By that a new title had to be invented for the paper, and since the word ceria did not belong to Mosander's vocabulary it became *On the new metals, Lanthanum and Didymium, which are associated with Cerium; and on Erbium and Terbium, new metals associated with Yttria* (1).

Berzelius was happy that Mosander was ready at last to announce his discoveries in a foreign language and wrote to his friend Carl Palmstedt in Gothenburg:

> "Mosander has put up a note on lanthan, didym and the constituents
> of yttria, which his brother-in-law major Beamish will bring to the
> meeting of the British Association at Cork in August; these
> discoveries will by that be presented to a big public and will then
> presumably be worked out by others, because for M. it goes too
> slowly to be completed by him in his life-time" (14; July 13, 1843).

The 1843 meeting of the British Association for the Advancement of Science was held at Cork because some weighty members of the Association "wished the light of science to be distributed as widely as possible through the benighted provinces" (17). The local arrangements at the meeting were as indifferent as the geologist Roderick Murchison had feared them to be, and long afterwards he remembered that "we never were so near shipwreck as at this Cork meeting" (17). In the long run, however, the three first papers, read in the opening session on August 18 of the section for chemistry at the Cork meeting, turned out to be more memorable than the indifferent local arrangements. Mosander's paper was first on the programme. When Beamish had read it, Thomas Andrews from Belfast presented one of his pioneer works within thermochemistry. In the paper *On the Heat of Combination* (18) he stated that when a base in a salt is displaced by another base, the heat of reaction depends only on the bases, being quite uninfluenced by the acid in the salt. Then the 25 year old James Prescott Joule from Salford, near Manchester, read his paper *On the Caloric Effects of Magneto-Electricity, and the Mechanical Value of Heat* (19). Experiments with a Voltaic pile and in the same circuit an electromagnetic machine, sometimes used as an engine, sometimes as a generator had convinced him that the principle of the perfect conservation of heat was indefensible. And by a dynamo-metrical apparatus attached to the electromagnetic machine he had ascertained that "a quantity of heat, capable of increasing the temperature of a pound of water by one degreee of Fahrenheit's scale, is equal to a mechanical force capable of raising a weight of about 838 pounds to the height of one foot". In contrast to Mosander's paper, however, Joule's paper was received with silence and disapproval at Cork.

The Cork paper remained Mosander's last - one can say only - publication on rare earths in spite of his final words in it: "I live in hopes that the knowledge already obtained will soon enable me to publish a more complete account of my investigations". Mosander was far from satisfied with his achievements, since he meant that he had just ascertained the existence of didymium oxide in ordinary cerium oxides and those of erbium and terbium in yttria, but he had not obtained any of them tolerably pure. During the following years he tried occasionally, when he had time for it, to find new methods to isolate them but his laboratory notes do not suggest that he ever succeeded. He had probably many times reason to lament in his talks with Berzelius about the "deviltries with which good results are so difficult to achieve" (11; January 12, 1847).

3.6. Notes

Note 1
Of the four children only the daughter Hulda Elisabeth Constance survived the parents. She married the Irish artillery officer Richard Pegott Beamish. Their daughter Huldine married the Swedish baron Carl Alexander Fock. Their daughter Carin (1888-1931) married in 1923 the German pilot, later on field marshal, Hermann Göring. Adolf Hitler liked her Swedish pea soup.

Note 2
Mosander like Berzelius used the formula $\overset{..}{T}e$ for titanic acid and $\overset{..}{S}n$ for stannic acid with dots above the atomic symbols of the metals, each denoting one oxygen atom.

Note 3
Mosander's formula for it was $2K\overline{C}y + Fe\overline{C}y$), since he used Berzelius's atomic weight from 1826 for potassium and used barred symbols to represent 'double'; $\overline{C}y$ is to be read $(CN)_2$.

Note 4
The reason why Lavoisier included the five known earths in his list of elements in 1789, but neither caustic soda nor potash, was presumbaly that Berthollet in 1785 had decomposed the volatile alkali ammonia in hydrogen and nitrogen which suggested that the two solid alkalis were compounds too.

Note 5
In 1815, from a mineral found in Falun, Berzelius isolated a substance that he thought was a new earth which he named thoria. Later on, however, he himself found that it was a basic yttrium phosphate. Instead he used the name thoria for the new earth that he isolated in 1829 from a mineral found at Brevik, Norway.

Note 6
Because of its electron configuration beryllium too is nowadays included among the alkaline earth metals and, for the same reason, also radium.

Note 7
Mosander's laboratory diaries "Mineral-analyser af C.G. Mosander" and "C.G. Mosanders Laborations-Bok" were donated to the Library of the Royal Swedish Academy of Sciences by his widow Hulda Mosander on April 13, 1859.

Note 8
What Erdmann thought was a new oxide was soon found to be a mixture of well-known oxides. Besides the mineral from which he had isolated that mixture Erdman had also analysed other minerals found at Brevik, Norway. One of them turned out to be a cerium mineral. In 1841 Erdmann named it *mosandrite* in Mosander's honour.

Note 9
Compt. Rend. Hebd. Séances Acad. Sci. 8 (1839) 356-357; *Ann. Mines*, Sér. 3, T 15 (1839) 422-423; *Phil. Mag.* 14 (1839) 390-391; *Arch. Pharm.* 18 (1839) 96; *Pogg. Ann.* 16 (1839) 648-649.

Note 10
The festival day of the Royal Swedish Academy of Sciences was, and still is, the 31st of March. In 1839, however, Easter Sunday fell on that date and the celebration of the centenary of the Academy was therefore postponed to Tuesday, April 2. The festival day of the Academy was the day when Berzelius 1821-47 presented his annual reports
"Årsberättelser" (Jahresberichte) on the progress in chemistry, and until 1838 also in physics, during the previous year.

Note 11
Pogg. Ann. 47 (1839) 207-210; also in *Ann. Pharm.* 32 (1839) 235-237.

Note 12
Annalen der Pharmacie, edited by Wöhler and Liebig, changed its name in 1840 to *Annalen der Chemie und Pharmacie*.

Note 13
With O = 16 Mosander would have assigned lanthanum the atomic weight 92.8 since he assigned lanthanum oxide the formula LaO.

Note 14
In accordance with his rules from 1818 for atomic symbols Berzelius used D as the symbol for the didymium atom. Mosander used in his note-book instead the symbol Di and so did most chemists after 1850.

Note 15
Scheerer announced the discovery of didymium in an appendix, *Nachschrift - Ein neues Metall, Didym, betreffend*, to the German translation in *Pogg. Ann.* of his paper, read at the Stockholm meeting (16). Wöhler announced it in a notice with the title *Ueber das Didymium, ein neues Metall* in *Ann. Chem. Pharm.* 44 (1842) 125-126.

3.7. References

1. Mosander, C. G., *On the new metals Lanthanium and Didymium, which are associated with Cerium; and on Erbium and Terbium, new [m]etals associated with Yttria*, Report of the Thirteenth Meeting of the British Association for the Advancement of Science held at Cork in August 1843, John Murray, Albermarle Street, London 1844, pp. 25-32. Also in *Phil. Mag.* 23 (1843) 241-254, *Pogg. Ann.* 60 (1843) 297-315, *Ann. Chem. Pharm.* 48 (1843) 210-223, *J. Prakt. Chem.* 30 (1843) 276-292, *Ann. Chim. Phys.* Sér. 3, T 11 (1844), 464-477, *Ann. Mines* Sér. 4, T 8 (1845) 157-177.

2. Mosander, C. G., *Något om Cer och Lanthan*, Förhandlingar vid de skandinaviske naturforskarnes tredje möte i Stockholm den 13-19 juli 1842, Stockholm, 1843, pp. 387-398.

3. Hevesy, G., *Chem. Rev.* 2 (1926) 1-41.

4. Söderbaum, H.G. *Jac. Berzelius Bref* 1- 6, utgivna av Kungl. Svenska Vetenskapsakademien; Almqvist & Wiksells Boktryckeri-A.B., Uppsala 1912-1932. Vol. 5, *Brevväxling med Nils Nordenskiöld (1817-1847)*.

5. Mosander C. G., *Kongl. Vetenskapsacademiens Handlingar* 1825 (1826) 204-210; German translation in *Pogg. Ann.* 6 (1826) 35-42.

6. Mosander, C. G., *Ibid.* 1825 (1826) 227-231; German translation in *Pogg. Ann.* 5 (1825) 501-506.

7. Mosander, C. G., *Ibid.* 1826 (1827) 299-310; German translations in *Arch. Gesammte Naturl.* 10 (1827) 470-483 and in *Pogg. Ann.* 11 (1827) 406-416.

8. Mosander, C. G., *Ibid.* 1829 (1830) 220-229; German translation in *Pogg. Ann.* 19 (1830) 211-221.

9. Mosander, C. G., *Ibid.* 1833 (1834) 199-226.

10. Berthier, P., *Ann. Chim. Phys.* 27 (1824) 19-28.

11. Wallach, O., *Briefwechsel zwischen J. Berzelius und F. Wöhler*, Vol. I (1823-1837), Vol. II (1838-1848), Verlag von Wilhelm Engelmann, Leipzig 1901.

12. Wöhler, F., *Pogg. Ann.* 13 (1828) 577-582.

13. Rose, H., *Pogg. Ann.* 59 (1843) 101-110.

14. Trofast, J., *Brevväxlingen mellan Jöns Jacob Berzelius och Carl Palmstedt* 1-3, utgiven av Kungl. Svenska Vetenskapsakademien, Lund 1979-1983, Vol. 3 (1839-1848).

15. Berzelius, J., *Tal vid öppnandet av Skandinaviska Naturforskare-Sällskapets första allmänna sammankomst i Stockholm*, Förhandlingar vid de skandinaviske naturforskarnes tredje möte i Stockholm den 13-19 juli 1842, Stockholm 1843, pp. 20-25.

16.   Scheerer, T., (a) *Chemisk Undersögelse af Gadoliniten fra Hitteröen og af et andet Mineral fra
         samme Findested;* (b) *Nogle chemisk-analytiske Erfaringer samlede ved Analysen af de naevnte
         Mineralier, Ibid.* pp. 373-386. A German translation of an extended version in *Pogg. Ann.* 56
         (1842) 479-505.

17.   Howarth, O. J. R., *The British Association for the the Advancement of Science: A Retrospect
         1831-1931*, Published by the Association at its Office in Burlington House, Piccadilly, W.1.,
         London 1931, p. 32.

18.   Andrews, T., Report of the Thirteenth Meeting of the British Association for the Advancement
         of Science held at Cork in August 1843, John Murray, Albemarle Street, London 1844, pp. 32-
         33.

19.   Joule, J. P., *Ibid.* p. 33.

CHAPTER 4

THE 50 YEARS FOLLOWING MOSANDER

F. SZABADVARY
Museum for Science and Technology
Budapest, Hungary H1500

C. EVANS
Ferguson Laboratory
University of Pittsburgh School of Medicine
Pittsburgh, PA USA 15261

4.1. Introduction

As we have seen in the previous chapter, the work of Mosander expanded the rare
earth family from two (Ce, Y) to six (Ce, La, Di, Tb, Er, Y) members. In the following
30 years, however, the subject ran out of steam. Berzelius had died and the rise of
organic chemistry eclipsed the popularity of inorganic chemistry. Moreover those rare
earths most readily susceptible to the traditional methods of discovery by fractional
precipitation and crystallisation had been identified. Further progress in this direction
required the development of improved techniques and new concepts. Thus it is no
accident that the next phase of rare earth discovery followed the introduction of
spectral analysis by Bunsen and Kirchhoff in 1859, and evolution of the notion that
elements could be organized in a logical fashion into predictable groups, as
exemplified by Mendeleev's periodic system of 1869 onwards. The former advance
provided a powerful new tool for seeking and identifying potentially novel elements,
while the latter gave an indication, albeit at the time an imperfect one, of the number of
additional rare earths that might yet exist.

4.2. The Terbium Dispute

Among the first to seize upon the new opportunities afforded by spectroscopy was the
Swiss chemist Jean Charles Marignac (1817-1894). Born in Geneva, Marignac studied
chemistry at the *École Polytechnique* in Paris, and then mineralogy and mining
engineering at the *École des Mines*. He returned to the *Academie de Genève* in 1841 to
teach chemistry. One of Marignac's major interests lay in determining precise atomic
weights of elements, including the rare earths. He calculated the atomic weight of
cerium, lanthanum and didymium as 47.26, 47.04 and 49.6, respectively (Marignac,

C. H. Evans (ed.), Episodes from the History of the Rare Earth Elements, 55–66.
© 1996 *Kluwer Academic Publishers. Printed in the Netherlands.*

1848; 1849). These values are in the correct relative order and were more accurate than determinations made by his contemporaries. In addition, Marignac made detailed studies of the chemical reactions of didymium and lanthanum (Marignac, 1853; 1855). Marignac is remembered for his discovery of ytterbium on the 22nd of October, 1878 (Marignac, 1878a; for a detailed discussion of this discovery see the following chapter in this book) and gadolinium (1880).

In 1864 another Swiss chemist, Marc Delafontaine (1837-1911) subjected erbium and terbium, whose discovery is described in the previous chapter, to close scrutiny. Although supporting the existence of erbium as a separate element, his data concerning terbium were equivocal (Delafontaine, 1864). Robert Bunsen (1811-1899), Professor of Chemistry at the University of Heidelberg, and his assistant Johan Bahr (1815-1875) reached a similar conclusion (Bahr and Bunsen, 1866). They immersed gadolinite in hydrochloric acid and, after removing the insoluble silica residue, precipitated with oxalic acid. The precipitate was dissolved in nitric acid, the ceria earths re-precipitated as sulphates, the filtrate re-precipitated with oxalic acid and this process continued until the didymium spectrum disappeared. The nitric acid solution, now containing only yttria earths, was evaporated to dryness, dissolved in water and allowed to crystallize. First to crystallize was the erbium salt, leaving yttrium in solution. From what was supposed to be terbium, Bahr and Bunsen (1866) could only spectroscopically detect erbium, yttrium and cerium. In later studies, Bunsen (1875) identified characteristic spectroscopic signals for erbium, yttrium, cerium, lanthanum and didymium, but failed to identify one for terbium.

Another important chemist to question the existence of terbium was Per Cleve (1840-1905) Professor of Chemistry at Uppsala University, Sweden. Moreover, Popp (1864) went one step further and denied the existence of both erbium and terbium. His spectroscopic analyses appeared to show that erbium is nothing more than didymium contaminated with cerium. Nevertheless, the existence of erbium was not seriously questioned by other chemists, especially following the detection of erbium in solar spectra by Young (1872).

The objections of Bunsen, Cleve and Popp notwithstanding, Delafontaine (1865) continued to support the point of view that terbium existed, but claimed differences from the terbium of Mosander which, he proposed, was actually identical to the erbium of Bahr and Bunsen. Delafontaine suggested that the element identified as terbium by Mosander should be re-named mosandrium. In later studies of the mineral samarskite, Delafontaine again detected mosandrium but, at the prompting of Marignac, reverted to the name terbium, and suggested that the original terbium which had been separated by Bunsen, should remain erbium (Delafontaine, 1877). The terbium and erbium of the present day retain this designation.

Jean Charles Marignac

Marc Delafontaine

Terbium's existence was subsequently confirmed from analyses of gadolinite by Marignac (1878 a,b). He calculated an atomic weight of 99 or 148.5 depending upon whether the formula of the oxide was TbO or Tb_2O_3. By a strange coincidence, terbium was simultaneously and independently isolated from samarskite by an American chemist, Lawrence Smith (1818-1883), who named it mosandrium (Smith, 1878). This name was discarded as a result of Delafontaine's victory in the ensuing priority dispute. The gathering momentum in favour of terbium's existence was further stimulated by Soret's report of the absorption spectrum of this element (Soret 1880). Definitive evidence came from publication of the exact, 194 line spectrum of terbium (Roscoe, 1882).

During his investigations of samarskite, Delafontaine had, in addition to terbium, identified a potentially novel element which he named philippium (Pp). Following the lead of Mendeleev, he predicted an atomic weight of 90-95 and suggested what some of its chemical properties might be (Delafontaine, 1878a). He later found another novel element decipium (Dp) with a projected atomic weight of 122 and a predicted absorption spectral line:λ =416 (Delafontaine, 1878b). Absorption spectra of both philippium and decipium were reported by Soret (1880). Nevertheless, philippium turned out to be illusionary. Roscoe (1882) showed that the formates of erbium and terbium form mixed crystals with the properties of philippium. Furthermore, the periodic table offered no space for such an element. Decipium, on the other hand, may have been what later became known as gadolinium.

4.3. Samarium and Others

At the same time as the terbium issue was being resolved, Delafontaine turned his attention to didymium. He was among the first to suggest that this presumed element was, in fact, a mixture. As we shall see, this turned out to be a prescient conclusion which required several more years for confirmation by Auer von Welsbach. (See Chapter 7). Delafontaine's opinion stemmed from the observation that the absorption spectrum of didymium isolated from samarskite differed from that isolated from cerite (Delafontaine, 1878b). Samarskite is a rare earth mineral discovered by G. Rose in the USA in 1839, but named after the Russian mining engineer V. Samarsky, who had found specimens of this ore in the Ural mountains during the 1860s. It was re-discovered in the USA in 1878 and immediately became the object of much attention as a rich source of potentially new rare earths.

The matter of didymium was taken up by Paul Lecoq de Boisbaudran (1838-1912), a French wine merchant who pursued chemistry as a hobby. Although Lecoq de Boisbaudran's results contradicted Delafontaine's on the matter of didymium spectra, unaccountable spectral lines seen by spark spectroscopy suggested the presence of a new element which, subsequent analysis confirmed, could be chemically separated from didymium and decipium. On the 16th of July, 1879 Lecoq de Boisbaudran announced the discovery of a new rare earth which he named samarium after the mineral samarskite from which it had been first obtained (Lecoq de Boisbaudran, 1879a,b).

Cleve (1885) soon lent his weight to the authenticity of this new element but the French chemist Eugene Demarçay (1852-1904) demurred, instead claiming, on the basis of absorption spectral lines, to have separated a new element from samarium (Demarçay, 1886). The matter remained unresolved for several years, even Lecoq de Boisbaudran the discoverer of samarium, later claiming to have observed the presence of additional rare earths in his element (Lecoq de Boisbaudran, 1892; 1893). On the other hand, Demarçay changed his mind and discarded his earlier opinion on the purity of samarium (Demarçay, 1893).

Such uncertainties led to intensive study of samarium. The result was the identification of two new lanthanides. One of these, europium, was finally purified by Demarçay (1896, 1901). Georges Urbain (1872-1930) subsequently purified europium from gadolinium, using a new method which used bismuth as a separating element. (Urbain and Lacombe, 1903; 1904).

In 1880 Marignac also turned to the examination of samarskite. Using fractional precipitation with potassium sulphate, followed by separation with oxalate, he obtained two potentially new rare earths, which he designated Ya and Yb (Marignac, 1880). Soret's spectral analyses suggested that Yb was samarium, but that Ya did not correspond to any known earth including Delafontaine's decipium (Soret, 1880). However, Delafontaine (1881) improved the purity of decipium and suggested that it did indeed correspond to the Ya of Marignac. It is likely that this was actually the case, but Marignac (1886) gave Ya the name gadolinium, which has remained to the present day.

4.4. The Division of Erbium

In confirming Delafontaine's conclusions on the existence of terbium, Marignac noted that the erbium precipitates he obtained were not homogeneous. Instead, there was a mixture of two elements. One which was pink with a characteristic absorption spectrum, retained the name erbium. The other was colourless and lacking an absorption spectrum. The two could be separated on the basis of treatment of the chloride with hyposulphurous acid. Under these conditions, the erbium salt precipitated, while that of the new element did not. Its atomic weight was predicted to be 115, if trivalent, or 172.6 if tetravalent. Recognizing this as an element intermediate in properties between yttrium and erbium, Marignac named it ytterbium (Marignac, 1878b). Soon afterwards, ytterbium was identified by Delafontaine (1878c) in a mineral from Virginia called sipylite. The Swede, Lars Nilson (1840-1899) repeated Marignac's separation of ytterbium, predicting atomic weights of 116 for a trivalent element or 174 in the case of tetravalency (Nilson, 1879).

Per Teodor Cleve (Courtesy Library, University of Uppsala)

But Nilson went further. In addition to ytterbium, he used a complicated fractionation procedure to obtain from gadolinite a novel element which he called scandium after Scandinavia. This new element possessed several unique spectral lines and was assigned an atomic weight of 80, in the case of trivalency (Nilson 1879), or 44, should it be tetravalent (Nilson, 1880). In either case, Nilson identified vacant spaces for his new element in Mendeleev's periodic table. Cleve (1879a) quickly confirmed the presence of scandium in gadolinite. From the mineral auxenite, Nilson later obtained pure ytterbium oxide and scandium oxide, assigning atomic weights of 173 to ytterbium and 44 to scandium (Nilson, 1880).

Nilson's discovery of scandium has a significance that goes beyond the identification of a new element. This is because the existence of such an element had been predicted by Mendeleev, on the basis of a vacant spot in his periodic table, in 1870. Mendeleev positioned this unknown element in the fourth period of the table, between calcium and titanium. On the basis of its assumed chemical similarity to boron, Mendeleev had named it eka-boron; after Nilson's work, it became re-named scandium. The confirmation of predictions such as these helped greatly to validate Mendeleev's system.

The division of erbium did not stop here. This time it was Cleve who doubted the homogeneity of what was now left of erbium after ytterbium and scandium had been removed. Three fractions were obtained. Subjected to spectroscopic scrutiny, these fractions led Cleve to propose the presence of two further elements, thulium and holmium. These names were derived from *Thule*, the Roman name for the northernmost region of the inhabitable world, and *Holmia* the Latin name for Stockholm. As it happened, the absorption spectrum of holmium had been detected earlier by Soret in a sample of erbium provided by Marignac. Soret (1878) had labelled this element X. These findings were subsequently confirmed by Lecoq de Boisbaudran (1879c). Thulium and holmium were assigned atomic weights of 113 and 108, respectively (Cleve, 1879b).

Holmium was later subjected to detailed study by Lecoq de Boisbaudran. He used a tedious method of separating rare earths from gadolinite, which involved 32 precipitations with ammonia and 26 with oxalate, followed by spectroscopic and fluorescence studies of the fractions thus isolated. But such careful labour was rewarded by the separation from holmium of a new earth, which he aptly named dysprosium after the Greek work *dysprositos* meaning "hard to get at" (Lecoq de Boisbaudran, 1886). Atypically for a new rare earth, discovery of dysprosium seems to have escaped the usual challenges surrounding such an event.

4.5. Separating the Twins

As described in the preceeding chapter, didymium was discovered by Mosander and named by him from the Greek word for twin. As we have seen, Delafontaine had suggested that didymium was not homogeneous. Evidence that this was indeed a double element was first provided by Auer von Welsbach (1858-1929). In addition to his virtuosity as a chemist, Auer von Welsbach was the first to realize the industrial uses of the lanthanides, as a result of which he became both famous and wealthy, and was made a baron.

Carl Auer von Welsbach was born in Vienna, the son of a director of the Imperial and Royal Court Printery. After attending school in Vienna, he studied chemistry at the University of Vienna with Professor Lieben. He subsequently attended the University of Heidelberg, studying chemistry under Bunsen and Kopp, but failed to complete his doctoral dissertation on the rare earths. However, while in Bunsen's laboratory he learned the techniques of spectral analysis. It is said that, during this time, Bunsen offered his opinion that "the future of gas lighting will be ensured by a solid glowing in the gas burner's flame". Auer brought rare earth minerals back with him to Vienna where he intended to complete his doctorate while working as an unpaid assistant to Professor Lieben.

While studying the separation of the lanthanide elements, he also began development of incandescent gas mantles based upon the abilities of various lanthanides to increase the light emission of a flame. In 1886 he announced his invention of his new gas lamp called an "actinophor". Commercial production of Auer's gas mantle required large volumes of rare earth minerals from which to extract the necessary elements. The search for suitable deposits led to the identification of large deposits of monazite sand in Brazil, which satisfied industrial demand. Until then, monazite had been used as ballast in sailing ships.

The commercial success of Auer's gas mantle made him independently wealthy. As a result, he was able to set up his own laboratory, first in a small basement, then via progressively larger premises to the pharmaceutical factory near Vienna that he bought in 1887. Such was Auer's fame and fortune that he was ennobled by the emperor Francis Joseph I, under which circumstances he procured castles and estates. In 1899 he moved to his castle near Rastenfeld in Austria, where he set up a research laboratory. His fame and influence may be gauged from the fact that a statue was erected in his honor and a picture of Auer von Welsbach was incorporated on the Austrian 20-Schilling note. It was in his castle laboratory that Auer invented a lighter flint based upon an alloy of cerium and iron. Commercialisation of this discovery only served to add to his wealth and reputation. Despite his creativity, Auer von Welsbach published only 16 papers. In later life, the importance of his gas lamps declined as they became progressively superceded by electric lighting. He died suddenly in 1929. A more detailed accounts of Auer von Welsbach's life and work are provided in chapter 7.

Let us now return to Auer's scientific studies of didymium. As we have seen, both Marignac and Delafontaine had questioned the homogeneity of this lanthanide. Marignac had gone so far as to invest considerable experimental effort into attempts to demonstrate the inhomogeneity of didymium, but he did not succeed (Marginac,

1853). However, such suspicions were rewarded by the discovery of samarium (Lecoq de Boisbaudran, 1879a,b) and thence europium (Demarçay, 1896) and gadolinium (Urbain and Lacombe, 1903).

Earlier doubts about the purity of didymium were increased by the atomic weight determinations of the Czech chemist Bohuslav Brauner (1855-1935), which showed variability in the values for didymium prepared from different sources. It is of interest to note that one of Brauner's motives here was to support the periodic system of his friend Mendeleev. Brauner aimed to obtain pentavalent didymium, as a result of which the problem of its placement in the periodic table would be simplified. He succeeded in isolating five fractions, three of which were mixtures of the other two. He proposed that didymium contained an additional component, termed Di-gamma (Brauner, 1882), a conclusion arrived at independently by Cleve (1882). Attempts to isolate Di-gamma by traditional fractional separation methods (Brauner, 1883) failed, as a result of which the credit for disentangling the components of didymium goes not to Brauner, but to Auer who had better success using a modified approach he had developed.

According to Auer's modification, nitrates were not decomposed by ignition, but by suspending the oxides in nitrate solutions. In this way, the precipitates were enriched in the less basic nitrates (Auer von Welsbach, 1883). Using this method, Auer succeeded in separating lanthanum from didymium by fractional crystallisation (Auer von Welsbach, 1884). Further application of the fractional crystallisation technique permitted the separation of two discrete fractions from didymium ammonium nitrate. This was laborious in the extreme, each crystallisation taking 1-2 days and over a hundred crystallisations being necessary for successful separation of the twins. Spectroscopic analyses confirmed that these were indeed different elements (Auer von Welsbach, 1885). Auer christened the green coloured element praseodidymium, from the Greek for green twin. The other element he called neodidymium (new twin). These names were later shortened to neodymium and praseodymium. In 1890, Bettendorf independently confirmed the existence of neodidymium and praseodidymium (Bettendorf, 1890). Despite Auer's discovery the name didymium persisted in the literature for many years, possibly because his findings were published in an obscure journal, the *Monatshefte für Chemie.*

4.6. Conclusions

During the period we have reviewed, roughly covering the last half of the nineteenth century, the number of rare earth elements jumped from six (Ce, La, Di, Tb, Er, Y) to fifteen (Ce, La, Pr, Nd, Sm, Eu, Gd, Tb, Er, Th, Ho, Dy, Yb, Y and Sc). Didymium had a long run for its money, but was no more. Missing from the modern list of rare earths were only lutetium and promethium. The discoveries of these two elements, and details of the ytterbium story, are described in the following two chapters.

4.7. References

Auer von Welsbach, C., 1883 Monatshefte für Chemie 4, 63.

Auer von Welsbach, C., 1884, Monatshefte für Chemie 5, 508.

Auer von Welsbach, C., 1885, Monatshefte für Chemie 6, 477.

Bahr, J. and R.W. Bunsen, 1866, Ann. 131,1;J für Prakt. Chem. 99, 274.

Bettendorf, A., 1890, Ann. 256, 159.

Brauner, B., 1882 Monatshefte für Chemie 3, 1, 486.

Brauner, B., 1883, J. Chem. Soc. 43, 278.

Bunsen, R.W., 1875, Pogg. Ann. 155,230,366.

Cleve, P.T., 1879a, C.R. Hebd. Séances Acad. Sci. 89, 419.

Cleve, P.T., 1879b, C.R. Hebd. Séances Acad. Sci. 89, 521.

Cleve, P.T., 1882, C.R. Hebd. Séances Acad. Sci. 94, 1528.

Cleve, P.T., 1885, Contributions to the Knowledge of Samarium.

Delafontaine, M., 1864, Arch. Phys. Nat. 21(2), 97; Ann. 134,99.

Delafontaine, M., 1865, Arch. Phys. Nat. 25,105; Bull. Soc. Chim. (Paris) 5(2),166.

Delafontaine, M., 1877, Arch. Phys. Nat. 59, 176.

Delafontaine, M., 1878a, Arch. Phys. Nat. 61, 273.

Delafontaine, M., 1878b, C.R. Hebd. Séances Acad. Sci. 87, 559.

Delafontaine, M., 1878c, C.R. Hebd. Séances Acad. Sci. 87, 634.

Delafontaine, M., 1881, C.R. Hebd. Séances Acad. Sci. 93,63.

Demarçay, E., 1886, C.R. Hebd. Séances Acad. Sci. 102,1551.

Demarçay, E., 1893, C.R. Hebd. Séances Acad. Sci. 117,163.

Demarçay, E., 1896, C.R. Hebd Séances Acad. Sci. 122,728.

Demarçay, E., 1901, C.R. Hebd. Séances Acad. Sci. 132, 1484.

Lecoq de Boisbaudran, P.E., 1879a, C.R. Hebd. Séances Acad. Sci. 88,322.

Lecoq de Boisbaudran, P.E., 1879b, C.R. Hebd. Séances Acad. Sci. 89,212; Arch. Phys. Nat. 2(3),119.

Lecoq de Boisbaudran, P.E., 1879c, C.R. Hebd. Séances Acad. Sci. 89,516.

Lecoq de Boisbaudran, P.E., 1886, C.R. Hebd. Séances Acad. Sci. 102,1003,1005.

Lecoq de Boisbaudran, P.E., 1892, C.R. Hebd. Séances Acad. Sci. 114,575.

Lecoq de Boisbaudran, P.E., 1893, C.R. Hebd. Séances Acad. Sci. 116,611,674.

Marignac, J.C., 1848, Arch. Phys. Nat. 8,265.

Marignac, J.C., 1849, Arch. Phys. Nat. 11,21.

Marignac, J.C., 1853, Ann. Chim. Phys. (Paris) 38(3),148.

Marignac, J.C., 1855, C.R. Hebd. Séances Acad. Sci. 42,288.

Marignac, J.C., 1878a, Arch. Phys. Nat. 61,283.

Marignac, J.C., 1878b, Arch. Phys. Nat. 64,87; C.R. Hebd. Séances Acad. Sci. 87,578.

Marignac, J.C., 1880, Arch. Phys. Nat. 3(3),413.

Marignac, J.C., 1886, C.R. Hebd. Séances Acad. Sci. 102,902.

Nilson, L.F., 1879, Ber. Dtsch. Chem. Ges. 12,551,554; C.R. Hebd. Séances Acad. Sci: 88,642; 91,118.

Nilson, L.F., 1880, C.R. Hebd. Séances Acad. Sci. 91,56,118.

Popp, O., 1864, Ann. 131,197.

Roscoe, H.E., 1882, J. Chem. Soc. 41,277.

Smith, L., 1878, C.R. Hebd. Séances Acad. Sci. 87,148.

Soret, J.L., 1878, Arch. Phys. Nat. 61,322; 63,89.

Soret, J.L., 1880, Arch. Phys. Nat. 4(3), 261.

Urbain, G. and H. Lacombe, 1903, C.R. Hebd. Séances Acad. Sci. 137, 792.

Urbain, G., and H. Lacombe, 1904, C.R. Hebd. Séances Acad. Sci. 138, 627.

Young, C.A., 1872, Am. J. Sci. 4(3), 353.

CHAPTER 5

ELEMENTS NO. 70, 71 AND 72:
DISCOVERIES AND CONTROVERSIES

HELGE KRAGH*
Roskilde University Centre
P. O. Box 260, 4000 Roskilde, Denmark

5.1. Introduction

The history of the discovery of the two last rare earth elements, ytterbium and lutetium, is a history of two priority disputes separated by a period of 16 years. The principal concern of the later and more bitter controversy was element 72, which is not a rare earth, but a zirconium homologue. Nonetheless, from a historical point of view the discovery of hafnium is an integral part of the discovery histories of the rare earths, which would be incomplete without hafnium. The two main contestants in the priority disputes, Georges Urbain and Carl Auer von Welsbach, were specialists in rare earth chemistry and highly regarded for their many contributions to this branch of chemistry. As an indication of their stature in the chemical community, both were nominated several times for a Nobel prize. Auer was nominated 10 times between 1918 and 1929, and Urbain 56 times between 1912 and 1936 (Crawford et al. 1987). All of Auer's nominations came from either Germans or Austrians, and almost all of Urbain's nominations were French. Their disagreements over the discoveries of elements did not, apparently, hurt their reputation. But it may well have contributed to the Swedish Nobel Committee's decision not to award either of them the valued award.

In this chapter, I outline the main events in the discovery histories of what after 1913 became known as the elements with atomic numbers 70, 71 and 72, that is, ytterbium, lutetium and hafnium. Special attention is paid to the controversies which arose over the discovery claims of these elements and other elements believed to populate the end part of the rare earths. The discovery of hafnium in 1923 emphasizes one important topic in the history of rare earth research, namely, the crucial importance of physical theory in the much discussed problem of the number and order of rare earth elements.

---

*Present Address:
Magnolieuangen 41,
3450 Alleroed, Denmark

C. H. Evans (ed.), Episodes from the History of the Rare Earth Elements, 67–89.
© 1996 *Kluwer Academic Publishers. Printed in the Netherlands.*

It was only after Niels Bohr had offered a rational foundation of atomic theory that the classification of the chemical elements, and that of the rare earths in particular, became a truly scientific question. This fact was not recognized by all chemists until much later, but today it furnishes a natural starting point for a historical periodisation of rare earth research. The discoveries of the ytterbium rare earths also illuminate the discovery process in general, adding support to the sociological view that discovery is primarily the attribution of status by the scientific community to certain claims (Brannigan 1981).

5.2. The ytterbium earths until 1905

When Marignac isolated the earth ytterbia from yttria in 1878, he concluded that he had discovered the oxide of a new element, which he proposed to name ytterbium. He assumed it to be trivalent and estimated its atomic weight to 172. Marignac's ytterbium was verified by Lars Fredrik Nilson and others and was generally accepted as a true element, but several chemists believed it to be a mixture of at least two elements. The discovery of ytterbium and Nilson's subsequent isolation of scandium from ytterbia were considered another triumph of Mendeleev's periodic system, which in its early versions predicted the existence of "eka-boron" with properties corresponding to those of scandium. Among the chemists who believed that ytterbium might possibly be a mixture were Georges Urbain in France and Carl Auer (von Welsbach) in Austria, who finally succeeded in splitting Marignac's ytterbium into two elements about 1907. But there were several earlier suggestions of ytterbium being a composite, most of them being pure speculations or based on slim evidence only. Thus, the two Viennese spectroscopists Franz Exner and Eduard Haschek investigated in 1899 a very pure sample of ytterbia they had received from Stockholm. Without explicitly saying that the spectrum revealed the existence of new elements, they stated that ytterbium was probably not "a unitary substance" (Exner and Haschek 1899, p. 1143). The following year, Eugène Anatole Demarçay claimed from examination of spectral lines the existence of a new element, called "Θ", between erbium and ytterbium. Demarçay suggested that it might possibly be part of Marignac's ytterbium (Demarçay 1900).

Also William Crookes in England believed that yttria contained more elements than generally assumed. He announced the discovery of several new rare earth elements and hypothesized that the chemical elements merely differed in their composition of "elementoids" or "meta-elements," a suggestion which has been seen as an anticipation of the notion of isotopy (Crookes 1888). It was not, however, speculations à la Crookes which characterized rare earth research in late nineteenth century, but painstaking experimental work over months or years in which theoretical ideas played little role. The period saw the announcement of many new rare earth elements, most of them believed to be parts of the accepted elements, but almost all of the claims were shown to be unfounded, the new elements being in fact mixtures of, or identical to, already known elements[1]. At the turn of the century, the following rare earth elements were accepted by most chemists: Lanthanum, cerium, praseodymium, neodymium, samarium, gadolinium, terbium, dysprosium, holmium, erbium, thulium, ytterbium, scandium and yttrium (see preceeding chapter). But there was much confusion as to the number of rare earth elements and as to whether all the mentioned elements were really elementary.

In retrospect, one cause for the confusion was the lack of a theory which could explain the periodic system of the elements in terms of atomic constitution. In lack of such theoretical knowledge, the periodic table was not, in itself, of much use to rare earth chemists. While most of the elements fitted well into Mendeleev's and Lothar Meyer's systems, the rare earths, with their very similar chemical properties, continued to be a puzzle in this respect. However, in spite of the lack of success in accounting for the position of the rare earths, the attempts to incorporate them in the periodic table was a source of inspiration to rare earth chemists and resulted in much new knowledge.

In Mendeleev's original classifications of 1869-71, the rare earths appeared indirectly as a separate block of 17 gaps, unconnected with the periodicity of the other elements. In the numerous versions of the periodic system which appeared in the following three decades, the earths appeared in almost as many ways. In some, they were considered as homologues of other elements; in others, as a special group or otherwise collectively placed[2]. Bohuslav Brauner, the Czech (Bohemian) chemist, worked for 30 years with the problem and proposed several classifications, none of them quite satisfactory, to incorporate the rare earths. In 1902 he grouped the rare earths collectively in group IV, but six years later he suggested dividing them between groups III to VIII (Brauner 1902, 1908). Periodic tables, in which the rare earths appeared in a single place, were proposed by, among others, Jan Willem Retgers in 1895 and Carl Benedicks in 1904 (Retgers 1895; Benedicks 1904).

A particularly interesting classification, which came to play some role in the later history of the rare earth elements, was the one with horizontal groups and vertical periods proposed by the Danish thermochemist Julius Thomsen in 1895[3]. The rare earth group was placed in the sixth period with 13 elements between cerium and an unknown element of atomic weight about 181. There was, furthermore, reserved place for another unknown element between neodymium and samarium, corresponding to the later promethium. Thomsen's system was not the first "inverted" system, with horizontal groups and vertical periods. Henry Bassett had suggested a somewhat similar system in 1892 and also he had placed the rare earths as a separate group, but with 18 elements (Bassett 1892).

The search for new elements included in Marignac's ytterbium cannot be sharply separated from other rare earth research in the period. During the 1880s and 1890s old methods of identification and separation were perfected and new ones introduced. The most important of the separation methods was fractionated crystallisation, a method based on the difference in solubility of corresponding salts of the various rare earths (Spencer 1919, pp. 21-56). Since the difference in solubility is only small, rare earth separation required the process to be repeated many times. This made rare earth chemistry laborious and time-consuming, a single separation sometimes requiring thousands of fractionations. For example, Urbain estimated to have performed, together with his assistants, a total of 20,000 fractionations over a period of 15 years (Perrin et al. 1939, p. 16).

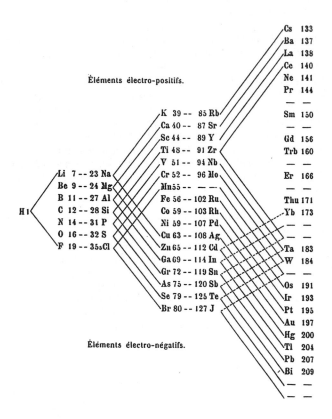

Figure 1 - Julius Thomsen's periodic system of 1895 (Thomsen, 1895)

5.3. Auer von Welsbach: aldebaranium and cassiopeium

Carl Auer von Welsbach studied under Adolf Lieben in Vienna and Robert Bunsen in Heidelberg, and started a research programme on the rare earths after graduation in 1880. A more complete biography of Auer is given in Chapter 7 of this volume. The programme did not only result in the splitting of didymia into two earth elements, praseodymium and neodymium, in 1885, but also in the profitable invention of the incandescent gas mantle and the automatic gas lighter. About 1900 Auer, who had then become a baron (*Freiherr*), began searching for the new element he believed was hiding in Marignac's ytterbium (D'Ans 1931). He started out with 500 kg of yttria earth from the *Auer Gesellschaft's* stock of monazite sand, and for separation he applied his favourite method, fractional precipitation of solutions of double ammonium oxalate in surplus ammonium oxalate. This method exploited analytically the fact that the solubility in ammonium oxalate of the rare earth oxalates varies systematically. After having separated erbium and thulium, he noticed in the residues a gradual change in the emission spectra of the ytterbium with further fractionation. In 1905 he had obtained a reasonably constant spectrum and by comparing it with the known spectrum of ytterbium he concluded that he had succeeded in splitting Marignac's ytterbium into two new elements. He first reported his conclusion in March 1905 to the Vienna Academy, where he said about ytterbium (Auer 1905):

> It consists mainly of two elements. Purification of both of the two
> new elements proceeds relatively easily, given the right choice of
> separation method. The spark spectra of the new elements are
> parts of the ytterbium spectrum, which can be considered as the
> sum of the two new spectra.

The word "mainly" (*hauptsächlich*) probably indicates that Auer was not sure that ytterbium was a composite of two elements only, and wanted to leave the door open for the future discovery of a third ytterbium element. The following year he gave a more detailed account of his results, which he claimed to provide "the first sure evidence of the compositeness of Yb" (Auer 1906a; he also referred briefly to his discovery claim in Auer 1906b, p. 464). At that time Auer seems to have believed not only that ytterbium was composite, but also that he was on his way to isolate "the true elements" from the other yttria earth elements as well: "After years of work, and with the support of numerous observations, I could draw the conclusion that almost all the elements of the yttria group until now known to chemistry are composite bodies and that it is possible to isolate the true elements" (Auer 1906a, p. 737).

Auer's publications of 1905-06 were of a preliminary nature. He promised soon to give a fuller report, which was the reason that in 1906 he only described his separation qualitatively and did not include data such as atomic weights or observed wavelengths. This he did only on December 19, 1907, when he presented a full report of his work to the Vienna Academy (Auer 1907). He admitted that he had not been able to separate the two ytterbium components completely, but found it unlikely that they should include a third element. Whereas at first he referred neutrally to the two components as

Yb I and Yb II, he now felt it appropriate to suggest new names and chemical symbols. Yb I, corresponding to Marignac's ytterbium, was called cassiopeium (Cp), and the new element, he had found therein, aldebaranium (Ad). More important than the names, Auer reported their atomic weights, which he had determined after Bunsen's method. For aldebaranium he found 172.90 and for cassiopeium 174.23. He also published his main evidence, the spark spectra of the two fractions, which he compared with the wavelengths of ytterbium earlier measured by the Swedish spectroscopist Tobias Thalén and more recently by Auer's compatriots Exner and Haschek (Thalen 1881; Exner and Haschek 1904). He found an almost complete agreement. However, at that time he was aware that his priority was questioned and that a Paris chemist, Georges Urbain, had recently claimed himself the discoverer of the two ytterbium elements.

5.4. Urbain: néo-ytterbium and lutecium

Georges Urbain was 35 years old when he announced his discovery of two new yttria earth elements and thus entered a controversy with Auer von Welsbach. A graduate from the *École de Physique et de Chimie* of 1894, Urbain specialized in rare earth chemistry in which field he wrote his doctoral dissertation and soon became an authority (Urbain 1898; Job 1939; Perrin et al. 1939). Urbain's study of the rare earths brought some order to the area by reducing the number of possible rare earth elements. In particular, he was able to refute Crookes's hypothesis of meta-elements and to show that only four of the many elements proposed to lie between samarium and holmium could be verified (Urbain 1925 a,b). Urbain had worked with fractionation of yttria earths since 1898, but only took up a study of the presumed composites of ytterbium about 1905. This work was part of a larger research programme and a continuation of earlier work on europium, gadolinium, terbium and dysprosium. His separation method did not differ essentially from Auer's, but Urbain's repeated fractionation was based on crystallisation of ytterbium nitrates rather than oxalates. He had perfected this method in early 1906 and then applied it to Marignac's ytterbium, obtained from the mineral xenotime, in order to determine whether its atomic weight was constant or not through the series of fractionations.

On November 4, 1907, Urbain presented his observations, based on a total of 800 fractionations, to the Paris Academy of Science. He announced that he had separated ytterbium into two new elements, for which he proposed the names neo-ytterbium (Ny) and lutecium (Lu). The first one was the principal component of ytterbium, and thus corresponded to Marignac's element in purified form. The new element, lutecium, was named after an old name for Paris, *Lutetia Parisorum*. Urbain's evidence consisted of roughly determined atomic weights of the elements and a table of 34 wavelengths for lutecium. He concluded that the reported observations showed that "Marignac's ytterbium is a mixture of two elements, neo-ytterbium and lutecium. The atomic weight of neo-ytterbium should not be too far from 170 and the atomic weight of lutecium not be much more than 174" (Urbain 1907). In the same paper he reported, as an additional result, that he had failed to confirm the existence of Demarçay's "Θ" element. That is, he had disproved it.

When Urbain read his paper to the Paris Academy, he was aware of Auer's earlier claim, but dismissed it on the ground that it lacked numerical documentation. Such documentation came 44 days later and at once initiated a bitter controversy over the priority of the discovery. Although the controversy took place only between Auer and Urbain, they were not the only ones who had separated ytterbium by 1907. Far away, at the University of New Hampshire, the British-American chemist Charles James worked with the same problem. In the summer of 1907, James had separated a large amount of what later would be known as element 71, probably by using fractional crystallisation of double magnesium ytterbium nitrates. He was about to publish his discovery when the news of Urbain's work reached him, which made him give up any claim and refrain from publishing his results. Had James been less cautious and less determined to verify his separation he might possibly have been recognized as the discoverer of the new element (Iddles 1930; Weeks and Leicester 1968, pp. 693-95).

5.5. The ytterbium controversy

When Urbain became aware of Auer's claim and documentation of December 1907, he hastened to reassert his own discovery claim in a supplementary paper to the Paris Academy (Urbain 1908a), and replied caustically to Auer's claim in German journals (Urbain 1908b). In the first paper, he provided more precise atomic weights, now determined to 171.7 (Yb) and 173.8 (Lu), and also reported the metals to be paramagnetic. Urbain did not question that Auer had separated ytterbium into two elements, but these, he said, were not new. They were simply lutecium and neoytterbium, which Auer thus had rediscovered rather than discovered. Then, as later, Urbain based his priority claim on the fact that he was first to announce publicly the numerical results which, they alone, made the discovery a fact and not merely a claim. Auer's publications of 1905-06 were descriptive and vague, without the crucial data. Hence, Urbain argued, they did not count. As to Auer's 1907 report - a detailed examination covering 45 pages - Urbain described it as "a brief note"! (Urbain 1908a).

Curiously, Auer maintained public silence about the matter throughout 1908. It was only after the verdict of The International Committee on Atomic Weights, published in January 1909, that he responded to what he felt was a grave injustice. The International Committee consisted until 1907 of Frank W. Clarke from the United States, Wilhelm Ostwald from Germany, Thomas E. Thorpe from England, and Henri Moissan from France. With the death of Moissan in 1907, a new member had to be appointed. The choice was Georges Urbain, who thus was in a central position to influence the decision in his own favour. At any rate, the Committee followed Urbain's arguments and concluded that "Urbain has clear priority [and] his nomenclature should be preferred" (Clarke et al. 1909). Lutecium entered the official table of chemical elements with atomic weight 174.0, whereas the lighter component appeared as "ytterbium (neoytterbium)" with atomic weight 172.0. The spelling of lutecium was later changed to lutetium. Incidentally, Urbain's original wish to provide neoytterbium with its own symbol (Ny) was not followed. It was decided to keep the old symbol for ytterbium, Yb.

The International Committee's decision in favour of Urbain made Auer "very angry" (*besonders verbittert*, according to D'Ans 1931, p. 84). In support of Auer's priority claim, the Viennese chemist Franz Wenzel announced in 1909 that Auer had earlier informed him by letter about the separation of what Auer at that time called "cassiopeum" and "aldebaranum"; the letter, dated June 5, 1906, reported the atomic weights $Cp = 174.28$ and $Ad = 172.52$ (Wenzel 1909). But this was at most an interesting historical detail that did not need to bother Urbain. After all, only public communications count in matters of scientific priority, and Auer published his atomic weights after Urbain.

Auer's own response to "the stubbornness with which Mr. Urbain has tried to claim the discovery of the ytterbium elements for himself," was read before the Vienna Academy on October 14, 1909 (Auer 1910). It was the response of an angry man. Auer knew that his claim would have to be based on his 1905-06 publications, which he therefore reviewed and reproduced in part. He believed that they gave "the scientifically exact proof" for the separation of ytterbium, and that their lack of wavelengths and atomic weights were of no importance. He had these data already in 1906 and kept no secret thereof, as shown by his letter to Wenzel. In what seems to be a desperate attempt to discredit his French rival, he intimated that Urbain had played foul. Without mentioning Urbain by name, he wrote (ibid., p. 509):

> Under the given circumstances, a chemist supplied with the necessary material, and making use of separation methods not too inappropriate, would face no difficulty in purifying the separation products of ytterbium to such purity that he would clearly recognize the change in intensity of the lines of the spark spectra; he would then be able to select the principal lines belonging to the individual elements.

Urbain, for one, did not miss the serious accusation implicit in Auer's message. In a reply of July 1910, written in German, he restated his point that Auer had failed to supply a "striking proof" in the form of numerical data before December 1907 (Urbain 1910). Auer's evidence, wrote Urbain, was of no more importance than the spectroscopically based suggestion of Exner and Haschek of 1899, which Auer apparently interpreted as a suggestion of ytterbium being composite (Exner and Haschek 1899). If Auer did not acknowledge Exner and Haschek as the discoverers of the composite nature of ytterbium, how could he, who did not report a single wavelength, possibly claim the priority for himself? As to Auer's intimation of foul play, Urbain was naturally upset. He noticed indignantly that Auer "goes as far as accusing me of simply plagiarizing him. ... It is disgraceful of Mr. Auer v. Welsbach to make such accusations against his colleagues and, at that, in an ambiguous way."

Although by 1910 Urbain was recognized as the discoverer of lutetium and (neo)ytterbium, and Auer's objections dismissed or ignored, this was not the last word in the priority controversy. Before dealing with the revised case, which only came up in 1923, let us review the situation as it was in 1910. Who was the "true" discoverer of the two elements? There is no doubt that Auer was the first to separate Marignac's ytterbium into reasonably pure components. Urbain and the International Committee granted him that much, but for them the crucial point was the lack of publicly available numerical data to confirm the claim. Urbain was the first to supply such data, although his atomic weight determinations were imprecise and of little value. Auer's determinations were earlier and more precise, but he failed to make them public in due time. Most likely, Auer would have been granted priority if he had published his wavelengths and atomic weights in, say, the spring of 1907. But he did not. With the International Committee's emphasis on publicly available, numerical documentation it is understandable that Urbain's priority claim was recognized. Yet the Committee's decision was far from objective, but based on an interpretation of the available data which predetermined the outcome of the controversy.

For example, the Committee chose to interpret Urbain's statements of the atomic weights - "not too far from 170" and "not much more than 174" - as determinations rather than mere estimates. Concerning the spectral evidence, Auer argued in his reply of 1909 that it was far from satisfying. Five of Urbain's lines were absent from Auer's much more complete spectrum of 1907, and four of Auer's most intensive lines felt in Urbain's spectrum. From this, Auer concluded that "Urbain ... has not succeeded in really separating ytterbium" and that the samples examined by his French rival were strongly polluted (Auer 1910). Apparently, the International Committee did not agree or did not care. But in 1923 it turned out that Auer had been essentially correct in his evaluation.

Priority disputes are negotionally settled according to social norms within the relevant scientific community. There plainly is no trans-social or trans-historical way to decide what person should be regarded the true discoverer. The late Rancke-Madsen proposed the fulfillment of the following two conditions to be sufficient and necessary for a person to be accepted as a discoverer of an element (Rancke-Madsen 1975):

> 1. He has observed the existence of a new substance which is different from earlier described substances, and this new substance is recognized by him or later by scientists as being elemental.
> 2. He must have published the discovery of the new substance in such a manner that it has been noticed by contemporaries outside the immediate circle of the discoverer.

According to these conditions, it would be reasonable to recognize Auer as the discoverer of lutetium or cassiopeium. But the conditions would not have been accepted by Urbain, who might have pointed out that they contain no mentioning of documentation or acceptance by the scientific community. They fail to discriminate between discovery claims and discoveries, or between claimed observations and scientifically valid observations.

5.6. Celtium

With respect to a theoretical understanding of the nature of the rare earths and their position in the periodic table, the discoveries of 1905-07 changed nothing. The existence of still more rare earth elements was a possibility which continued to attract the interest of many chemists, Auer and Urbain included. In 1907 Auer began a search for a possible third ytterbium constituent, but decided four years later that aldebaranium and cassiopeium could not be further separated. This conclusion was later confirmed by means of X-ray spectroscopy of samples of his two elements (Friman 1916). Auer did however believe to have separated thulium into two new elements in 1911, and the Austrian spectroscopist Josef Maria Eder claimed in 1916 to have identified two new elements in Auer's fractions. Auer agreed and suggested to Eder to name the elements "denebium" and "dubhium." Four years later Eder claimed yet another element, which he called "welsium" in honour of Auer von Welsbach (Eder 1923; Figurowski 1981, p. 261). None of the claims were taken seriously by either chemists or physicists and were soon forgotten. Urbain's candidate for a new rare earth seemed to fare better. The French chemist investigated the band spectrum of his neoytterbium and observed that it varied in an unpredictable manner. He first mentioned the possibility of a third ytterbium element in an oral communication to the French Physical Society on July 3, 1908 (Urbain 1909b, p. 172), and in 1909 he wrote that "I am inclined to consider it [neoytterbium], too, a mixture of two elements" (Urbain 1909a).

At that time Urbain had started a research programme on the magnetic properties of the rare earths which he used as an additional means of identification. Urbain found the magnetic susceptibility to be an additive property and devised methods to calculate the relative composition of a binary mixture by measuring the magnetic susceptibility of the mixture and its constituents. After further fractionation of lutetium, he obtained a mother-liquor which refused to crystallise and with spectral and magnetic properties indicating the presence of a new element. In 1911 he claimed to have discovered a new rare earth element, which he called celtium (Ct) in honour of his fatherland (Urbain 1911). He believed the element to be next to lutetium in the periodic system, that is, number 72. As proof of his discovery, he published 24 spectral lines not formerly assigned to any rare earth, and reported a steady fall in the magnetic susceptibility through the refinements. He measured the molar magnetic susceptibility of the highly refined fraction, believed to be celtium, to be about 1.6. However, although appearing in part of the chemical literature, celtium failed to win general acceptance[4]. Urbain seems himself to have been somewhat uncertain about his claim, which was based on a very small amount of the substance. He assumed his sample of celtium not to be pure and was unable to determine its atomic weight.

Shortly after the announcement of celtium, atomic physics experienced a revolution which promised to create order in the rare earths where chemists had only found chaos. The means for an understanding of the rare earth group, and the periodic system in general, were Niels Bohr's atomic theory and Henry G. J. Moseley's method of X-ray spectroscopy which built upon the new model of the atom. According to Moseley, the position of an element in the periodic system could be uniquely determined by its atomic number, which revealed itself through $\sqrt{v/R}$, where v is the frequency of the characteristic X-ray emitted by the element and R is Rydberg's constant. Moseley's successful application of X-ray spectroscopy for chemical purposes made Urbain seek his support with regard to the confirmation of celtium[5]. Shortly before the outbreak of the Great War, Moseley examined Urbain's sample of celtium. Because of the war and Moseley's untimely death - he was killed during the Gallipoli campaign in 1915 - his results remained unpublished. Urbain chose to interpret Moseley's examination as indecisive: although it did not support celtium, neither did it refute his discovery. He wrote to Rutherford in September 1915 that "Celtium has not been revealed; nevertheless there remains a place for celtium in Moseley's series between lutecium and tungsten" (Heimann 1967). In 1922 he repeated that Moseley's experiments did not contradict the existence of celtium (Urbain 1922). Urbain's interpretation was biased by his wish to preserve celtium, for in fact Moseley left no doubt that the examined sample of celtium contained no element of atomic number 72. According to an account of Moseley's unpublished report to the British Association meeting in Sydney in August 1914, "the X-ray spectrum indicates that it [celtium] is a mixture of previously known earths" (Nature 1914). And in a letter to Rutherford, written shortly after the experiments, Moseley said (Heilbron 1974, p. 236):

> Celtium has proved most disappointing. I can find no X ray spectrum in it other than those of Lutecium and Neoytterbium, and so Number 72 is still vacant. My own impression is that the very definite spectrum (visible) given by "Celtium" is a secondary Lutecium spectrum, which is masked in what Urbain calls "Lutecium" by 50% of Neoytterbium, while "Celtium" is Lutecium with only a small &age of Ny.

After the war, when X-ray spectroscopy was further developed, Urbain teamed up with Alexandre Dauvillier, a young X-ray specialist at Maurice de Broglie's laboratory in Paris. Dauvillier measured the L spectrum of a mixture of lutetium and ytterbium oxides and found two very weak lines with wavelengths agreeing with those of $Z = 72$ in a Moseley diagram (Dauvillier 1922). With this result, Urbain considered celtium definitely confirmed. "It is now unquestionable," he wrote in May 1922, "that the element of atomic number 72 is actually celtium ... [which] has conclusively won its place among the chemical elements" (Urbain 1922, p. 1349).

Dauvillier and Urbain's rediscovery of celtium was for a time widely accepted, both by chemists and physicists. Rutherford arranged an English translation of Urbain's article and publicly endorsed the discovery (Rutherford 1922). In Copenhagen, Bohr was disturbed, but at first tended to accept the discovery. However, Bohr had reasons to believe that element number 72 was not a rare earth at all, and late in August 1922 he had decided that celtium was probably a misinterpretation.

5.7. Hafnium

Bohr's reasons were based on his new atomic theory of the periodic system which he worked out between 1921 and 1923 (Kragh 1979). In this theory, energy levels were classified as n_k, where n is the principal and k the azimuthal quantum number (k is one unit larger than the azimuthal quantum number of quantum mechanics). The rare earths were characterized by the building up of the 4_4 level in the N shell, whereas the two outer levels (O and P shells) remained unchanged. Since lutetium would contain the maximum of 4x8 N electrons, the next element would be a zirconium homologue and not a rare earth. Bohr's theoretical conclusions with regard to the end of the rare earth group received some support from new X-ray experiments, but these were too inexact to decide clearly whether Z = 72 was a rare earth or not. At any rate, when Bohr delivered his Nobel address in December 1922, he was convinced that the element of atomic number 72 had to belong to group IV. Urbain and Dauvillier's claim was, he stated, "incompatible with the conditions of the quantum theory" (Bohr 1922, p. 42).

The periodic system which grew out of Bohr's analysis was a modified version of Thomsen's 1895 system, which Bohr knew well and the inspiration of which he acknowledged. Bohr used the system in his Nobel Lecture on December 11, 1922, where element 71 appeared as Cp. Forty days earlier, he had used the same system, but then designated the element as Lu (Bohr and Coster 1922). In Bohr's system, sometimes called the Bohr-Thomsen system, the building up of the 4_4 quantum level was marked by a box which ended with ytterbium, the last element with an incomplete 4_4 level. The placing of Z = 71 outside the box may lead to the wrong impression that there are only 13 rare earth elements in Bohr's system. However, this was only a result of Bohr's wish to arrange the elements according to principles of atomic theory.

In order to settle the matter about element 72, Bohr needed a specialist in X-ray spectroscopy as well as pure samples of rare earths. The first he obtained from Lund, where the Dutch physicist Dirk Coster worked at Manne Siegbahn's laboratory. The necessary rare earths he obtained from Auer, who had also provided Siegbahn with samples. In Copenhagen, Coster teamed up with George Hevesy, Bohr's friend and collaborator, and the search for element 72 started. The idea to look for the element in zirconium minerals seems to have come from Fritz Paneth, who did not believe in celtium. In a review article on the periodic system from the summer of 1922, Paneth placed element 72 in group IV and recommended to look for what he called "eka-zirconium" in zirconium minerals (Paneth 1922, p. 383). This Hevesy and Coster did, with the result that they could claim the discovery of a new group IV element - hafnium (Hf) - on January 2, 1923 (Coster and Hevesy 1923a). Incidentally, Bohr and his collaborators originally decided to call the new element "danium". It was an editorial misunderstanding which made "hafnium" appear in *Nature* and subsequently become the accepted name (Kragh and Robertson 1979).

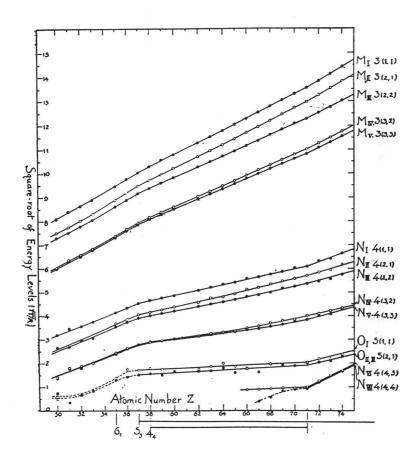

Figure 2 - The rare earths according to X-ray spectroscopy. This Moseley diagram, published by the Japanese physicist Yoshio Nishina in 1925, provided the first experimental proof that the rare earth group ends with element 71 and comprises 14 elements (Nishina 1925). The measurements were made at Bohr's institute in Copenhagen in 1924, where Nishina worked with Bohr and Coster. In the rare earths, the 4_4 level is filled up with electrons, initiated in Z=58 and completed in Z=71. According to Bohr's theory, the binding energy of the 4_4 electrons is approximately constant throughout the rare earth group. The end of the group will thus appear as a sharp upward bend.

The announcement of hafnium was, at the same time, a denouncement of celtium. It led to a bitter priority dispute which involved not only the discoverers in Paris and Copenhagen, but also Alexander Scott in London. The latter, mineralogist and chief chemist at the British Museum, believed for a short time to have discovered element 72 - which he called "oceanium" - in 1918 (Scott 1923). I shall not here follow the London-Copenhagen and Paris-Copenhagen controversies, which I have detailed elsewhere (Kragh 1980), but only mention a few of the high points.

Urbain and Dauvillier at first maintained that Coster and Hevesy had merely detected the rare earth celtium in zirconium minerals, but were gradually forced to admit that number 72 was in fact a group IV element. Had Urbain not believed so strongly that number 72 was a rare earth he might have discovered the element himself in 1922. For in that year he analysed a mineral from Madagascar very rich in hafnium, but only reported the existence of zirconium, half of which may actually have been hafnium (Urbain and Boulanger 1922). As it was, Dauvillier and Urbain now argued that although Urbain's 1911 celtium claim was perhaps a mistake, Dauvillier's observations of X-rays still remained conclusive proof that they had discovered celtium in May 1922. Unfortunately for Urbain and Dauvillier, they were unable to produce good X-ray spectra of celtium or otherwise provide documentation for their claim, which rested on the two exceedingly faint lines reported in words only in 1922. During 1923 the controversy escalated, fuelled by professional jealousy, personal disagreements, and the tense political situation between Germany and France at the time. Although no Germans participated in the discovery of hafnium, in France and England it was widely seen as a teutonic conspiracy. By the end of 1924, Coster, Hevesy and others had prepared large amounts of pure hafnium and obtained an impressive knowledge of its chemistry and physical properties (Hevesy 1925, 1927; Rose 1926). Celtium remained two elusive X-ray lines and a name in French periodic tables. In spite of the weak evidence in favour of celtium, the International Committee on Chemical Elements (with Urbain as its chairman) refused throughout the 1920s to give official credit to hafnium. Instead the Committee accepted two names and symbols, Ct and Hf, for the same element, a rather exceptional decision. Not less exceptional, neutrality was secured in the Committee's table of elements in 1925 simply by omitting number 72 altogether (JACS 1925).

As an unexpected result of the celtium-hafnium dispute, new light was thrown on the old controversy between Auer and Urbain concerning the discovery of ytterbium's constituents.

5.8. New light on old elements

The establishment of hafnium as element 72 naturally raised the question of the true nature of celtium. How could Urbain in 1911 have discovered a chemical element for which there was no place in the periodic system? How could he have found 24 new spectral lines in a component of lutetium and yet failed to observe these in his lutetium spectrum of 1907? If celtium was not identical with hafnium, what was it?

In April 1923 Hans M. Hansen and Sven Werner, two physicists at Bohr's institute in Copenhagen, subjected Auer's cassiopeium to new spectroscopic investigation and also reexamined the earlier cassiopeium spectra reported by Exner and Haschek in 1911 and Eder in 1915 (Hansen and Werner 1923; Exner and Haschek 1911; Eder 1915). Comparing the optical spectrum of cassiopeium with Urbain's 1911 celtium spectrum they found that all the celtium lines belonged to Auer's element, whence they concluded that the celtium of 1911 had just been element 71. But, since Urbain had reported his celtium lines as new and not detected them in his lutetium spectrum of 1907, this indicated that the 1907 spectrum had only contained few lines belonging to element 71. The conclusion was inevitable: Urbain's lutetium of 1907 had only contained very little of element 71, which element he had only isolated four years later, under the name celtium.

Hansen and Werner's conclusion received further support from an examination of the elements' magnetic properties undertaken by Coster and Hevesy (Coster and Hevesy 1923b). Urbain did not report absolute values of the magnetic susceptibilities of his neoytterbium and lutetium, but found the ratio Yb/Lu = ca. 4.1 for the oxides Yb_2O_3 and Lu_2O_3. In 1911, he measured the molar magnetic susceptibility of celtium to be 1.6 and stated that it was 3 to 4 times smaller than that of lutetium. Auer's elements had been examined by the Austrian physicist Stefan Meyer in 1908, who found the ratio Ad/Cp = ca. 4.8 (Meyer 1908). Considering these data, Coster and Hevesy inferred that the magnetic susceptibility of Urbain's 1907 lutetium had been about 5.5 and thus considerably larger than Auer's cassiopeium. Since the paramagnetism decreases with the refinement of the ytterbium fractions (pure element 71 is diamagnetic), in 1907 lutecium had been much less pure than cassiopeium.

The Copenhageners' reinterpretation of the events from 1907-11 resulted in a campaign to reinstate Auer as the discoverer of element 71. On March 27, 1923, Coster informed Auer about their intentions (Kragh 1980, p. 287):

> There can be given only one reason why the name proposed by Urbain for element 71 is used, namely that this name happens to have been accepted by both the "International" and the German Atomic Weight Commission. But it is a question if this reason is sufficient, and we will do what we can to put the matter on the agenda again. We will take care that the problem of elements 70, 71 and 72 will soon be discussed in the German journals.

Although Bohr did not himself enter the controversy, behind the scene he was active in the pro-cassiopeium campaign. He supported Auer's claim in letters to the Austrian chemist, and to Rutherford he wrote in August 1923[6]:

> My argument for changing the name [from lutetium to cassiopeium] was not only that Coster and Hevesy had shown that in contrast to Welsbach [sic!] who from the first has had very pure preparations of the element, Urbain has up to the present moment only had a smaller fraction of this in his preparations,

what was the reason for his wrong conclusion as regards the
properties of the element 72. Before all however was the fixation
of the atomic number under consideration made by Siegbahn on
the basis of very pure specimens prepared by Welsbach what was
brought with it, that the name Cp from Siegbahn's paper, has got
over in most recent textbooks of atomic theory...

As mentioned by Bohr, most physics publications changed from lutetium to
cassiopeium, thereby providing support to Auer's priority claim. As a result of the
Copenhagen physicists' campaign to reconsider Auer's priority, the question was taken
up by several German and Austrian chemists. For example, Paneth demanded celtium
and lutetium to be abandoned and declared the International Committee's decision to
ignore Auer's claim "completely unjustified" (Paneth 1923, p. 172). At that time
German science was boycotted by the former allied nations and Germany excluded
from The International Committee on Chemical Elements. The German counterpart,
the *Deutsche Atomgewichtskommission*, chaired by Otto Hönigschmid, decided in
November 1923 to give full credit to Auer's discovery of element 71 which
consequently was named cassiopeium. Element 70 was still called ytterbium "as a
tribute to the memory of Marignac," and of course celtium was dismissed and hafnium
accepted. The argument in favour of cassiopeium, signed by Hönigschmid, Otto Hahn,
Max Bodenstein and Richard Meyer, was as follows (Hönigschmid et al. 1924):

> [We] must assign the priority for the splitting of Marignac's
> ytterbium to Auer v. Welsbach, for his elaborate publications on
> ytterbium's compositeness confirm his claim of the years 1905
> and 1906; furthermore, he had obviously proceeded far beyond
> Urbain with the purification of both ytterbium components at the
> time of Urbain's first publication.

However, in spite of the strong German (and Scandinavian) support of cassiopeium,
lutetium continued to be used in most chemical literature and Urbain to be recognized
as the discoverer of element 71. In 1930 the now reorganized International Committee
accepted hafnium as the sole name for element 72, but kept lutetium for element 71
(JACS 1931).

Brauner, the rare earth veteran, followed the controversy with particular interest. He
had never been a friend of Auer, who, Brauner intimated, had unfairly capitalized on
his work on didymium in 1882 to turn himself into the discoverer of neodymium and
praseodymium three years later. Brauner sided with Urbain in the dispute and accused
Coster and Hevesy to have stolen the discovery of element 72 from Urbain and
Dauvillier[7]. Urbain, on his side, recognized Brauner as the real discoverer of
neodymium and praseodymium and saw in the unfortunate fate of celtium an analogy
to the history of the splitting of didymium (Urbain 1925b).

The new light thrown on the 1905-11 discoveries by the Copenhagen physicists
provides an interesting example of how contemporary science may intervene in the
history of science and change its records. Such intervention is by no means exceptional

and is particularly common in the history of chemistry (Kragh 1987, pp. 159-67). It is noteworthy that the critical examination of the historical data performed by Hevesy and his collaborators did not rely on new knowledge or methods particular to the 1920s. The examination of the spectral data and the reported magnetic properties could easily have been made in, say, 1915, and would then have resulted in a similar conclusion. But in 1915, when the cassiopeium-lutetium dispute was long forgotten and lutetium and cassiopeium believed to be just two names for the same substance, there would have been no reason to undertake such an examination. Scientists are forward-looking and usually find the study of the past a waste of time. It was only under an external pressure, the controversy over element 72, that the Copenhagen scientists turned to "historical" studies.

The conclusion of the German Atomic Weight Commission in 1924 seems inescapable. Whereas it was reasonable enough to assign Urbain the priority of element 71 in 1909, it was no longer so in 1923. Neither is it today, of course. The fact that Urbain's priority rests on false premises, viz., that his 1907 sample was reasonably rich in element 71, has not really prevailed in the chemical and historical literature, though. In most works, whether written by chemists or historians, the "international" version of the discovery of element 71 is taken for granted and Auer only mentioned as an independent re-discoverer of the same element. For example, Weeks and Leicester uncritically repeat the "international" version and even fail to mention that the splitting of ytterbium was the subject of a controversy (Weeks and Leicester 1968, pp. 692-96). And according to Partington's *History of Chemistry*, "the last rare-earth element, lutetium, was discovered by Urbain and independently by Auer von Welsbach" (Partington 1972, p. 910).

With the discovery of hafnium, the search for new rare earths was limited to the still not discovered element 61. Auer, among others, hoped to find this element and had some of his rare earths specimens examined by means of X-ray spectroscopy in Copenhagen. However, no lines indicating the presence of $Z = 61$ turned up. Continued search for the element in fractions of neodymium and samarium nitrates was equally unsuccessful. Optical examination revealed no new element and the disappointing result was confirmed X-spectroscopically by Ludwig Prandtl in Munich (Auer 1926). Neither this failure nor the physicists' assurance that no new rare earth element except no. 61 could possibly exist, kept Auer from continuing his search for new elements. To him, the unpredictiveness of traditional rare earth separation was a fascination that no physical theory could destroy. He loved "the romance of the search for elements according to the old methods" and continued to look for new elements in thulium fractions until two days before his death (D'Ans 1931, p. 85).

5.9. Conclusion

The discovery of lutetium and ytterbium as components of Marignac's ytterbium earth was structured by a long tradition in rare earth chemistry which made the attribution of discovery unproblematic. Although not unexpected, the discovery was neither predicted nor predictable. None of the earlier suggestions of ytterbium being composite were scientifically based. Neither they nor periodic tables with a vacant place following Marignac's ytterbium (such as Thomsen's) can reasonably be called predictions. They were inspired guesses, but lacked the rational foundation which characterises true predictions.

The controversy which followed the claims of 1907 was not about whether or not the new elements had in fact been discovered. It was unanimously agreed that this was the case. The subject of the dispute was one of priority only. The official attribution of priority to Urbain depended on the criteria used by the International Committee and its interpretation of the evidence supplied by Auer and Urbain. With the chosen criteria and the chronology of the publications, Urbain was awarded priority. The criteria were of course not objective or beyond dispute, but a choice which could well have been different and then having made Auer the discoverer.

The subsequent discovery claim of celtium was very different. It was not, either in 1911 or in 1922, universally endorsed. The transformation of a discovery claim into a discovery requires the acceptance of the scientific community. Between 1911 and 1923 celtium existed as a "semi-discovered" element, accepted by many but not all chemists. In this case the controversy was about the content of the claim, i.e., the existence of celtium, and not about who had first discovered it (although Urbain sought to bring up that subject in 1923). The fact that the Lu/Cp controversy was about priority and the Ct/Hf controversy about the validity of scientific claims, helps explain the intensity of the latter dispute. Urbain's scientific reputation was at stake, and the political circumstances of the post-war years magnified the controversy so it became a matter of national honour as well. Both controversies were epistemic and internalist in that they contained only arguments recognized to be scientific and took place within the scientific communities. But, as in other cases of controversy, they included a social aspect in the sense that they were between groups with opposing social interests. This element was much clearer in the Hf/Ct case, whereas the earlier dispute between Auer and Urbain mostly involved internal social reasons. On the whole, the many cases of controversies over the discovery of chemical elements, whether rare earths or not, constitute an important source for the general study of scientific controversies (Brante and Elzinga 1990).

Although controversies over priority may seem foolish and irrational, they are often scientifically productive because they lead to research that might not have been conducted in the absence of a controversy. The competitive pressure forces the involved scientists to launch arguments and counter-arguments, the effect of which will be a deepened understanding of the discovery. This feature is nicely illustrated by the cases here examined. Much of the early work on ytterbium and lutetium was related to the controversy between Urbain and Auer. In the case of hafnium, the rapid accumulation of knowledge of that element in 1923, and the reexamination of the Lu/Cp case, were direct outcomes of the controversy.

One last aspect deserves mentioning, viz., the role played by atomic physics in the discoveries, and the somewhat strained relationship between the two sister sciences, chemistry and physics. Although rare earth chemistry utilized physical techniques such as optical spectroscopy and magnetic measurements, it relied predominantly upon traditional chemical skills. Many of the practitioners in this arch-chemical branch of science distrusted physical theory and advanced physical methods. Even more than Urbain, Auer was an old-line chemist who neither liked nor understood the intrusion of atomic physics on chemistry. The tension between traditional chemistry and the new physics played no role in the Lu/Cp case, but it was an important element in the Ct/Hf controversy, which can be seen as a boundary conflict between professional interests. It concerned, among other matters, which of the parties was the real expert on element identification, with the adjoining claims of social status and authority. Many chemists felt it intolerable that outsiders, young physicists with hardly any chemical knowledge at all, confidently claimed to settle an important chemical problem. This feeling was clearly expressed by Brauner, who was frustrated over the atomic physicists' easy solution of the problem he had fought with for 40 years, the position of the rare earths in the periodic system. In 1923, he accused the Copenhagen physicists of lacking the "chemical sense" which was crucial in rare earth research and could only be obtained by years of hard labour in the laboratory (Brauner 1923). But neither Brauner's nor other chemists' objections could hide the fact that the new physics had changed the basis of rare earth research and stripped it of its chemical romance.

5.10. Notes

[1] See the list of seventy reported rare earth elements (1794-1920) in Mellor 1924, pp. 504-05, and Karpenko 1980 for a comprehensive list of spurious element discoveries.

[2] For references to and details on the history of rare earths in the periodic system, see Rudorf 1904 and Van Spronsen 1969.

[3] Thomsen 1895. For the historical context of Thomsen's system, see Kragh 1982. See also Hevesy 1927, pp. 5-7.

[4] In James Spencer's textbook of 1919, probably the most authoritative book on rare earth chemistry in the early part of the century, celtium entered as a fully accepted element. The book shows that in 1919 X-ray spectroscopy was still unknown to, or unappreciated by, most chemists. Among the methods described for identifying rare earths, X-ray spectroscopy was not mentioned (Spencer 1919).

[5] For further historical details, see Heimann 1967, Kragh 1979, pp. 178-81, and Mel'nikov 1983.

[6] Bohr to Rutherford, 22 August 1923 (Bohr Scientific Correspondence). Reproduced from a handwritten copy with the permission of the Niels Bohr Archive, Copenhagen.

[7] Brauner 1923. For the intimation of plagiarism, see the letter from Brauner to Max Speter of 18 May, 1933, as translated in Weeks and Leicester 1968, p. 689.

5.11. References

Auer von Welsbach, C. (1905), "Vorläufiger Bericht über die Zerlegung des Ytterbiums in seine Elemente," *Anzeiger der mathematisch-naturwissenschaftliche Klasse der kaiserliche Akademie der Wissenschaften 42,* 122.

Auer von Welsbach, C. (1906a), "Über die Elemente der Yttergruppe," *Sitzungsberichte der mathematisch-naturwissenschaftliche Klasse der kaiserliche Akademie der Wissenschaften 115,* IIb, 737-47.

Auer von Welsbach, C. (1906b), "Bemerkungen über die Anwendung der Funkenspectren bei Homogenitätsprüfungen," [Liebig's] *Annalen der Chemie* 351, 458-66

Auer von Welsbach, C. (1907), "Die Zerlegung des Ytterbiums in seine Elemente," *Sitzungsberichte der mathematisch-naturwissenschaftliche Klasse der kaiserlische Akademie der Wissenschaften 116,* IIb, 1425-69.

Auer von Welsbach, C. (1910), "Zur Zerlegung des Ytterbiums," *Sitzungsberichte der mathematisch-naturwissenschaftliche Klasse der kaiserlische Akademie der Wissenschaften 118,* IIb, 507-12.

Auer von Welsbach, C. (1926), "Über einige Versuche zur Auffindung des Elementes Nr. 61," *Chemiker-Zeitung 50,* 990.

Bassett, H. (1892), "A tabular expression of the periodic relations of the elements," *Chemical News 65,* 3-4, 19.

Benedicks, C. (1904), "Über die Atomvolumina der seltenen Erden und deren Bedeutung für das periodische System," *Zeitschrift für anorganische Chemie* 39, 41-48.

Bohr, N. (1922), "The structure of the atom," pp. 7-44 in *Nobel Lectures, Physics,* 1922-1941 (Amsterdam: Elsevier, 1965).

Bohr, N. and Coster, D. (1922), "Röntgenspektren und periodisches System der Elemente," Zeitschrift für Physik 12, 342-37.

Brannigan, A. (1981), *The Social Basis of Scientific Discoveries* (Cambridge: Cambridge University Press).

Brante, T. and Elzinga, A. (1990), "Towards a theory of scientific controversies," *Science Studies* 3: 2, 33-46.

Brauner, B. (1902), "Über die Stellung der Elemente der seltenen Erden im periodischen System von Mendelejeff," *Zeitschrift für anorganische Chemie 32,* 1-30.

Brauner, B. (1908), "Über die Stellung der Elemente der seltenen Erden im periodischen System," *Zeitschrift für Elektrochemie 14,* 525-28.

Brauner, B. (1923), "Hafnium or celtium," *Chemistry & Industry 42,* 884-85.

Clarke, F. W. et al. (1909), "Report of the International Committee on Atomic Weights, 1909," *Journal of the American Chemical Society 31,* 1-6.

Coster, D. and Hevesy, G. (1923a), "On the missing element of atomic number 72," *Nature* 111, 79.

Coster, D. and Hevesy, G. (1923b), "On celtium and hafnium," *Nature 111,* 462-63.

Crawford, E., Heilbron, J. L. and Ullrich, R. (1987), *The Nobel Population 1901-1937* (Berkeley: Office for History of Science and Technology).

Crookes, W. (1888), "Elements and meta-elements," *Journal of the Chemical Society 53,* 487-504.

D'Ans, J. (1931), "Carl Freiherr Auer von Welsbach," *Berichte der deutsche chemische Gesellschaft 64,* 59-92.

Dauvillier, A. (1922), "Sur les séries L du lutécium et de l'ytterbium et sur l'identification du celtium avec l'élément de nombre atomique 72," *Comptes Rendus 174*, 1347-49.

Demarçay, E. A. (1900), "Les spectres du samarium et du gadolinium," *Comptes Rendus* 131, 387-90.

Eder, J. M. (1915), "Der Bogenspektrum des Cassiopeiums, Aldebaraniums, Erbiums und des in weitere Elemente gespaltenen Thuliums," *Sitzungsberichte der mathematisch-naturwissenschaftliche Klasse der kaiserliche Akademie der Wissenschaften 124,* IIa, 707-828.

Eder, J. M. (1923), "Die Spektralanalyse der seltenen Erden," *Annalen der Physik 71,* 12-18.

Exner, F. and Haschek, E. (1899), "Über die ultravioletten Funkenspektra der Elemente," *Sitzungsberichte der mathematisch-naturwissenschaftliche Klasse der kaiserliche Akademie der Wissenschaften 108,* IIa, 1123-51.

Exner, F. and Haschek, E. (1904), *Wellenlängetabellen für Spektralanalytische Untersuchungen auf Grund der Ultravioletten Bogenspektren der Elemente* (Vienna: F. Deuticke).

Exner, F. and Haschek, E. (1911), *Spektraltafeln* (Vienna: F. Deuticke).

Figurowski, N. (1981), *Die Entdeckung der chemischen Elemente und der Ursprung ihrer Namen* (Köln: Aulis Verlag).

Friman, E. (1916), "On the high-frequency spectra (L-series) of the elements lutecium-zinc," *Philosophical Magazine 32,* 497.

Hansen, H. M. and Werner, S. (1923), "On Urbain's celtium lines," *Nature 111,* 461.

Heilbron, J. L. (1974), *H. G. J. Moseley. The Life and Letters of an English Physicist 1887-1915* (Berkeley: University of California Press).

Heimann, P. M. (1967), "Moseley and celtium: The search for a missing element," *Annals of Science 23,* 249-60.

Hevesy, G. (1925), "Recherches sur les propriétés du Hafnium," *Kongelige Danske Videnskabernes Selskab, Meddelelser,* VI:7, 1-147.

Hevesy, G. (1927a), *Die Seltenen Erden vom Standpunkte des Atombaues* (Berlin: Julius Springer).

Hevesy, G. (1927b), *Das Element Hafnium* (Berlin: Julius Springer).

Hönigschmid, O. et al. (1924), "Vierter Bericht der Deutschen Atomgewichtskommission," *Berichte der deutschen chemische Gesellschaft 57* B, I-XXXVI.

Iddles, H. A. (1930), "The Charles James Hall of Chemistry of the University of New Hampshire," *Journal of Chemical Education 7,* 812-20.

JACS (1925), *Journal of the American Chemical Society 47,* 597.

JACS (1931), *Journal of the American Chemical Society 53,* 1627.

Job, P. (1939), "Notice sur la Vie et les Travaux de Georges Urbain," *Bulletin de la Societé Chimique de France* 6, 744-66.

Karpenko, V. (1980), "The discovery of supposed new elements: two centuries of errors," *Ambix 27,* 77-102.

Kragh, H. (1979), "Niels Bohr's second atomic theory," *Historical Studies in the Physical Sciences 10*, 123-86.

Kragh, H. (1980), "Anatomy of a priority conflict: The case of element 72," *Centaurus 23*, 275-301.

Kragh, H. (1982), "Julius Thomsen and 19th-century speculations on the complexity of atoms," *Annals of Science 39*, 37-60.

Kragh, H. and Robertson, P. (1979), "on the discovery of element 72," *Journal of Chemical Education 56*, 456-59.

Kragh, H. (1987), *An Introduction to the Historiography of Science* (Cambridge: Cambridge University Press).

Mellor, J. W. (1924), *A Comprehensive Treatise on Inorganic and Theoretical Chemistry*, Vol. 5 (London: Longmans, Green & Co.)

Mel'nikov, V. P. (1983), "Some details in the prehistory of the discovery of element 72," *Centaurus 26*, 317-22.

Meyer, S. (1908), "Magnetisiergszahlen seltener Erden," *Sitzungsberichte der mathematisch-naturwissenschaftliche Klasse der kaiserliche Akademie der Wissenschaften 117,* IIa, 955-61.

Nature (1914), *Nature , 94* ,353.

Nishina, Y. (1925), "On the L-absorption spectra of the elements from Sn(50) to W(74) and their relation to the atomic constitution," *Philosophical Magazine 49*, 521-37.

Paneth, F. (1922), "Das periodische System der chemischen Elemente," *Ergebnisse der exakten Naturwissenschaften 1*, 362-402.

Paneth, F. (1923), "Über das Element 72 (Hafnium)," *Ergebnisse der exakten Naturwissenschaften 2*, 163-76.

Partington, J. R. (1972), *A History of Chemistry*, Vol. 4 (London: Macmillan).

Perrin, J. et al. (1939), *Hommage à Georges Urbain* (Paris: Hermann & Cie.).

Rancke-Madsen, E. (1975), "The discovery of an element," *Centaurus 19*, 299-313.

Retgers, J. W. (1895), "Beiträge zur Kenntnis des Isomorphismus, XI," *Zeitschrift für physikalische Chemie 16*, 577-658.

Rose, H. (1926), *Das Hafnium* (Braunschweig: Vieweg & Sohn).

Rudorf, G. (1904), *Das periodische System. Seine Geschichte und Bedeutung für die chemische Systematik* (Hamburg: L. Voss).

Rutherford, E. (1922), "Identification of a missing element," *Nature 109*, 781.

Scott, A. (1923), "Isolation of the oxide of a new element," *Journal of the Chemical Society, Transactions 123,* 311-12.

Spencer, J. F. (1919), *The Metals of the Rare Earths* (London: Longmans, Green and Co.).

Thalén, T. R. (1881), "Sur les raies brillantes spectrales des métaux scandium, ytterbium, erbium et thulium,"*Öfversigt af Kungliga Vetenskaps Akademiens Handlingar 38*, no. 6, 13-21.

Thomsen, J. (1895), "Classifications des corps simples," *Kongelige Danske Videnskabernes Selskab, Oversigt,* 132-36.

Urbain, G. (1898), *Recherches sur la Séparation des Terres Rares* (University of Paris dissertation).

Urbain, G. (1907), "Un nouvel élément: le lutécium, résultant du dédoublement de l'ytterbium de Marignac," *Comptes Rendus 145*, 759-62.

Urbain, G. (1908a), "Sur le lutécium et le néoytterbium," *Comptes Rendus 146*, 406-08.

Urbain, G. (1908b), "Zur Zerlegung des Ytterbiums in seine Komponenten," *Chemiker-Zeitung 32*, 730.

Urbain, G. (1909a), "Europium, gadolinium, terbium, dysprosium, neoytterbium und lutetium," *Chemiker-Zeitung 33*, 745-46.

Urbain, G. (1909b), "Révision des poids atomiques des terres rares," *Bulletin de la Société Chimique de France 5*, 133-72.

Urbain, G. (1910), "Lutetium and neoytterbium or cassiopeium and aldebaranium," *Zeitschrift für anorganische Chemie 68*, 236-42.

Urbain, G. (1911), "Sur un nouvel élément qui accompagne le lutécium et le scandium dans les terres de gadolinite: le celtium," *Comptes Rendus 152*, 141-43.

Urbain, G. (1922), "Les numéros atomiques du néo-ytterbium, du lutécium et du celtium," *Comptes Rendus 174*, 1349-51.

Urbain, G. and Boulanger, C. (1922), "Sur la composition et les caractères chimiques de la thortveitite de Madagascar," *Comptes Rendus 174*, 1442-43.

Urbain, G. (1925a), "Twenty five years of research on the yttrium earths," *Chemical Reviews 1*, 143-85.

Urbain, G. (1925b), "Discours sur les éléments chimiques et les atomes. Hommage au Professeur Bohuslav Brauner," *Recuil des Travaux Chimiques des Pays-Bas 44*, 281-95.

Van Spronsen, J. W. (1969), *The Periodic System of Chemical Elements: A History of the First Hundred Years* (Amsterdam: Elsevier).

Weeks, M. E. and Leicester, H. M. (1968), *Discovery of the Elements* (7th edn., Easton, Pa.: Journal of Chemical Education).

Wenzel, F. (1909), "Zur Spaltung des Ytterbiums," *Zeitschrift für anorganische Chemie 64*, 119-20.

CHAPTER 6

THE SEARCH FOR ELEMENT 61

JACOB A. MARINSKY
Department of Chemistry
State University of New York at Buffalo
Buffalo, NY 14214 USA

6.1. Introduction

The Oak Ridge National Laboratory was established during the Second World War. Its primary objective, as a part of the Manhattan Project in 1943, was the development, as rapidly as possible, of a feasible process for the production of the fissionable nuclides, Pu^{240} and U^{235}, at a purity level and rate sufficient for the assembly of the nuclear bomb which was under development concurrently at the laboratory complex established in the University of California Project at Los Alamos, New Mexico (1,2) as another part of the Manhattan Project. In the Oak Ridge, Tennessee complex the Clinton Laboratories, identified as X-10 at the time (1943), were operated by the University of Chicago. It was involved with the development of a chemical process for the separation of slow-neutron fissionable Pu^{240}, produced in the graphite-moderated nuclear reactor built for this purpose, from the components of the fuel elements and the radioactive products formed simultaneously. Research programs to develop a technique for the separation of the slow-neutron fissionable U^{235} isotope from its much more abundant naturally occurring isotope, U^{238}, were carried out in two other installations in Oak Ridge. The first of these, known as Y-12, was operated by Tennessee Eastman and examined electromagnetic techniques for their separation. Union Carbide was in charge of operations at K-25, the second installation used to examine the feasibility of uranium isotope separation by a gaseous diffusion process. In this process advantage was taken of mass-based differences in the diffusion of $U^{235}F_6$ and $U^{238}F_6$ through porous barriers.

In the spring of 1942 a section of the Metallurgical Laboratory at the University of Chicago, also a component of the Manhattan Project, was placed under the direction of Charles D. Coryell. Its designated function was radiochemical research of the nuclear fission process. Its objective was to be the separation and identification of more than thirty fission product elements together with the characterization of their many radioisotopes (3). In late summer of 1943 Charles D. Coryell and several members of his "nuclear chemistry" group moved into the newly constructed Clinton Laboratories.

91

C. H. Evans (ed.), Episodes from the History of the Rare Earth Elements, 91–107.
© 1996 Kluwer Academic Publishers. Printed in the Netherlands.

The rest of his section remained at the Metallurgical Laboratory under the direction of Nathan Sugarman.

The graphite-moderated reactor was used for neutron irradiation of uranium to provide the path to investigation of the complex array of radioisotopes associated with the decay of the fission product elements formed. The most difficult of these to separate, identify, and characterize were the rare earths (4). Because of their great chemical similarity, such assessment of the "praseodymium group" (Pr, Nd, and element 61) was particularly inaccessible. Because the discovery of element 61 in nature was subject to question (5-14), its separation, identification, and characterization in the rare earth fission product fractions provided by these studies became an especially important objective.

The search for element 61 in nature had begun with Moseley's demonstration that it should occur between neodymium and samarium in the periodic table (15). The best known of such efforts were initiated by Harris, Hopkins, and Yntema (16) and Rolla and Fernandez (17) who sought to separate element 61 for identification from naturally occurring rare earth mixtures by fractional crystallization. Both groups of investigators claimed its discovery in nature in 1926. Hopkins, Harris, and Yntema based their claim of discovery on the occurrence of 130 otherwise unidentified lines common to the arc spectra of the neodymium- and samarium-enriched sequence of fractionally crystallized rare earth samples (18,19). Changes in the solution absorption spectrum compiled in the course of the rare earth separation process were claimed as further evidence for the presence of element 61 as well. Finally, lines of the L x-ray spectra of several of the presumably element 61-enriched samples were found to be in agreement with the lines predicted for element 61 by Moseley's rules. On the basis of these observations the name "illinium" (after the state and university) was proposed for element 61.

The independent and prior discovery of this element was claimed in the same year by Rolla and Fernandez on the basis of spectral data similar to those of Hopkins and his coworkers. They proposed the name "florentium" for the element (17).

These claims were disputed vigorously by Prandtl (5-7) who stated that he was able to reproduce the fading of the neodymium absorption lines and the appearance of new lines, as reported by Hopkins, Harris and Yntema, by changing the proportions of pure neodymium and samarium. He also pointed out that contamination by barium, chromium, platinum and bromine could give rise to the x-ray lines observed by Hopkins, Harris, and Yntema. Further support for this view was provided by Mattauch's rule, formulated from examination of the pattern of occurrence of the stable nuclides (20), which suggested that the existence of a stable isotope of element 61 is highly unlikely. In addition, the fact that Hopkins and Rolla and Fernandez failed to provide further evidence for the existence of element 61 in nature after announcing new programs for element 61 isolation in greater quantity (21,22) was felt to be indicative of the error in their claim of discovery. The lack of success of the many investigators who have searched for this element (7-14) was significant as well.

Early in 1943 Nathan Ballou, before leaving Chicago for Oak Ridge, discovered a low energy, beta-emitting fission product. His studies at X-10 showed that it decayed with a half-life of 3 to 4 years (23). Another rare earth activity produced in fission was the 11-day emitter (24) discovered by Harrison Davies (25) at X-10 in 1944. Parallel studies were carried out in the Metallurgical Laboratory at Chicago (3).
It was possible, at the time, to perform procedures which narrowed the assignment of the 11 day and long-lived activities to neodymium or element 61. These procedures, developed concurrently, included oxidative fusion with KOH (26, 29) and digestion in K_2CO_3 solutions (26, 30-33). They effected fractional separation of lanthanum and praseodymium from neodymium (27-32).

Early in 1944 the Army Specialized Training Program that I was a part of was eliminated. The war was not going well, and army personnel was needed elsewhere. Late in March 1944 I was one of eighteen soldier students at Purdue University selected from a class of at least 1000 to be transferred to Oak Ridge upon dissolution of the Army Specialized Training Program (ASTP). An examination given to qualified enlisted personnel, while I was assigned as a private to teach Weight and Balance Control in Aircraft to Air Force cadets at Chanute Field in Rantoul, Illinois, had led to my acceptance in this program. When I left Purdue University upon the dissolution of the ASTP in March 1944 I needed only one additional semester of study to add a bachelor's degree in Chemical Engineering to the one in Chemistry that I had received in 1939 from the University of Buffalo.

Upon our arrival in Oak Ridge, Tennessee we were assigned to the Special Engineering Detachment (SED) that had been established by the military as one answer to the scientist shortage that prevailed at that time. We joined a number of other soldiers in the pool that had been established to facilitate our placement in the various unfilled laboratory positions. After being interviewed by Charles D. Coryell and W. H. Sullivan, his assistant, I, together with other soldiers and a few civilians, was assigned to X-10 (Clinton Laboratories). That day, April 4, 1944 (4-4-44), it was my good fortune to be afforded the opportunity to initiate active participation in the development of a new and rapidly expanding area of science. I remember clearly the startling revelation by L. Riordan, who was in charge of security at X-10, of the science fiction-like military objectives of the Laboratory that I had just joined. Professor Coryell, after Riordan's revelations, then led us into the high security area of the Laboratory. With his shirt tails flying he pointed to a black, barnlike building on a hill overlooking the research laboratories and offices. As he identified the nuclear reactor that would be the focal point of our research activities he could not suppress the excitement he felt.

A well organized introduction to radiochemical concepts as well as instruction in the proper handling of radioactivity was then provided our group as soon as we reached the research laboratories. The effective transfer of knowledge essential to our proper indoctrination was accomplished in a teaching program developed and led by Dr. David N. Hume.

I recall my apprehensive reaction to the earlier selection of members of the group by the various Section Chiefs of X-10. I was the last member of the group to be assigned but the luckiest. I had been chosen to join Professor Coryell's research program.

With my assignment to this program I was able to have my wife Ruth join me in Oak Ridge. One room in a dormitory (M-6), built in the center of the Oak Ridge community to accommodate married military personnel, provided housing. Because of the shroud of secrecy that had to be maintained by all Oak Ridge personnel, her first words to me when she finally arrived in Oak Ridge were "My God, in your letters and phone calls you sounded like Agent X-9" (a secret agent in a comic strip that was popular at the time). I surreptitiously covered my mouth with my hand and whispered "X-10".

In the laboratory there was sizable interaction with the Section Chiefs and the Group Leaders. Discussion of experiments in progress often led to their contribution of ideas and even active participation in the experiments. To add to the level of excitement there were seminars and lectures by the superstars of science (Wheeler, Teller, Nordheim, Wigner). It was an exhilarating experience for me.

6.2. Separations and Identifications

Shortly after Ruth's arrival I was assigned the problem of identifying a several-day gamma emitter in the rare earth fission product fraction. With the cooperation of my more experienced colleagues, especially L.E. Glendenin and N. Ballou, I entered a program that first employed the classical fusion techniques employed in the earlier investigations by Ballou (26-33). In the course of this research, fusion with sodium nitrate (26,34) showed that the 11d gamma emitter was an isotope of praseodymium, neodymium, or element 61 (29). Fusion with potassium hydroxide then showed it could not be an isotope of praseodymium, which narrowed the identification of the isotope to neodymium or element 61 (29).

In the midst of these studies I learned of the research of Adamson, Swartout and associates who were the first to demonstrate the applicability of the ion-exchange technique to the separation of trace elements from each other (35). Investigations of fundamental aspects of the ion-exchange process, conducted separately by Adamson (36) and Schubert (37), then provided further insight with respect to the element separation potential provided by this approach. Cohn, Tompkins and coworkers, in their application of the method to the large scale separation of fission products, successfully isolated the rare earth isotopes from the other fission product elements (38). Because I was perceived as the rare earth specialist in the Clinton Laboratories as a result of my active research of rare earth fractionation procedures, my participation in an assessment of the rare earth separation that had been effected was sought by Cohn and Tompkins. The degree of fractionation of the rare earth elements and the order of their elution was determined for this purpose through analysis of yttrium, the "praseodymium group" (praseodymium, neodymium and element 61) and cerium in the column effluent samples collected in the course of the rare earth fractionation (39). The graphical representation of these results presented in Figure 1 showed that the praseodymium group was eluted more slowly than yttrium and at an appreciably faster rate than cerium when five percent (by weight) citric acid solution, brought to a pH of approximately 2.9 with concentrated ammonium hydroxide, was used to elute the rare earth elements from the Amberlite IR-1 resin column. One could conclude that the 11d γ-emitting activity in the "praseodymium group" eluted more rapidly than the β activity in the "praseodymium group" from a comparison of the γ to β ratio in successive effluent fractions that is listed in Table 1. This result was consistent with the earlier assignment of the 11d γ emitter to neodymium or element 61 that was based on its fractionation from 13.8d praseodymium in the earlier KOH fusion studies.

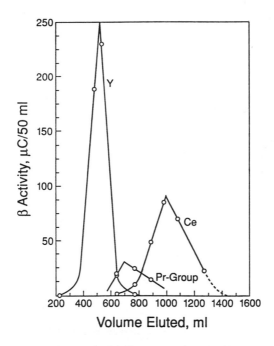

Figure 1   Elution curves of rare earth fission activities

Table 1 - Differential β and γ Elution Rates in the Praseodymium-group Activities

Eluate cut, ml	γ activity, c/m/ml	β activity, c/m/ml	γ to β activity ratio x10^4
560 - 725	1.1 x 10^3	7.75 x 10^6	1.4
725 - 828	0.26 x 10^3	8.20 x 10^6	0.3
828 - 948	0.10 x 10^3	4.85 x 10^6	0.2

To supplement these observations a series of experiments was then designed to elucidate further the order of elution of the rare earths. Such information was essential for the unambiguous identification of the isotopes of element 61 and neodymium that are formed in the fission process. I realized that with such knowledge firmly established, a definitive experiment that would lead to the positive identification of element 61 would for the first time, become feasible.

Carrier-free cerium, yttrium and lanthanum fission product activities were employed in these experiments carried out together with Khym (39). The radioactive isotopes of praseodymium and europium, made by slow neutron irradiation (n, γ) of the stable elements, were used with less than 10 mg of the element added as carrier. In these experiments the radioactive isotopes were adsorbed individually at the top of 10 ml columns of Amberlite IR-1 resin. Each activity was then eluted with the buffered citrate solution at various pH values. The results of this study are presented in Fig 2 by plotting percent activity removal versus effluent volume. The pH of influent and effluent at a particular effluent volume is also provided in the figure.

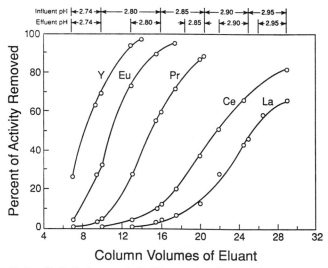

Figure 2 Elution of individual rare earths by 5% ammonium citrate solution

In another experiment macro-quantities of samarium, europium, and neodymium were adsorbed on and eluted from an IR-1 resin column with five percent ammonium citrate at a pH of 2.75. A spectrographic analysis of the fractions collected showed that europium, eluted first, was followed by samarium and then by neodymium.

These experiments showed that the order of elution of the rare earths from a cation exchange resin, using five percent ammonium citrate at a pH of approximately 2.8 to 2.9 as the eluant, is that of decreasing atomic number (39). This adsorption, elution cycle became the basis of the next experiment that was performed to identify radioactive isotopes of element 61 and neodymium that are formed in the fission process (40).

In one set of experiments not mentioned so far, europium had appeared to be eluted from the resin column at a faster rate than yttrium. To remove this ambiguity a sample in which both yttrium and europium activities were present was subjected to the adsorption and elution cycle. A rare earth fraction, chemically isolated from irradiated uranium that had decayed for 1.5 years after a 126 day irradiation, was adsorbed on and eluted from an Amberlite IR-1 resin column with five percent ammonium citrate buffer solution at a pH of 2.6. Activity measurements of the 57d Y^{91} and the 2y Eu^{155} in the effluent samples showed that yttrium was eluted more rapidly than the europium (39).

Such corroboration of the results obtained for the separate yttrium and europium adsorption and elution experiments summarized in Fig 1 showed that the yttrium behaves like a rare earth in the gadolinium region. This resemblance in chemical behavior to rare earth elements of higher atomic number in the adsorption and elution cycle is consistent with the parallel separation tendencies of yttrium and the higher atomic number rare earths in the fractional precipitation studies (30-32).

No questions remained at this point with respect to the path that should be taken for positive identification of element 61. The fact that the rare earth elements had been conclusively shown to be eluted from a cation-exchange resin column in the reverse order of atomic number, and the fact that even neighboring rare earths had been demonstrated to be rather effectively separated in the elution process, dictated the design of this definitive experiment. To prepare a suitable source a rare earth fission product sample was isolated chemically from a uranium fission product mixture on the 15 mg of lanthanum carrier that was added. Its cerium activity content was then effectively removed by the precipitation of $Ce(IO_3)_4$ after oxidation of the cerium carrier added for this purpose to the 4+ state. Most of the yttrium, europium and samarium activity in the sample was then removed from the praseodymium group (praseodymium, neodymium, and element 61) sought for study by precipitation of the lanthanum carrier as the very dense, fine crystalline precipitate of the mixed carbonate $K_5(La, Pr, Nd, element\ 61)_3(CO_3)_4$ formed during digestion of the mixture in concentrated potassium carbonate. Effective removal of yttrium, europium and samarium in the supernatent was assured by repeating the precipitation step, which provides incomplete separation, several times.

Before I could proceed with adsorption of this sample at the top of the Amberlite IR-1 cation exchange resin column, made ready for elution with buffered five percent ammonium citrate I was, to my dismay, shifted to Dr. G.E. Boyd's Section to assist in improvement of the process that had been developed for the isolation of Ba^{140}-La^{140} from the fission products. Improved yield of carrier free Ba^{140} in the process was essential to provide the 40h La^{140} daughter at the radioactive concentration level needed for its effective use in the development of a trigger for the atomic bomb under development at the Los Alamos laboratory.

After several months, an interval that felt much longer, I was able to return to the rare earth identification experiment (40). The "praseodymium group" source, prepared earlier, was adsorbed at the top of the Amberlite IR-1 column. Elution with five percent ammonium citrate at a pH of 2.75 was initiated. Fractions of eluant were collected, and the beta and gamma activities in each fraction were measured. These data plotted versus the volume of eluant provided the elution curve shown in Figure 3 (29,40). Four beta-activity peaks were resolved. The gamma activity measured was found to be associated with the third beta-activity peak.

Absorption studies of the beta activity associated with the first peak assigned the radio-activity to 57d ^{91}Y. The fact that this peak is so much smaller than the other peaks indicates that fractionation of ^{91}Y from the lanthanum carrier-isolated rare earth fission product fraction in the series of potassium carbonate digestions performed, had been quite as effective as projected. Study of the adsorption and decay characteristics of the fourth peak showed its activity to be due solely to 13.8d Pr^{143}. The second and third prominent beta activity peaks occurring between the first (^{91}Y) and fourth (Pr^{143}) peaks had to be due to element 61 (41) and neodymium (element 60) (29), the elution order of the rare earths having been shown to be the reverse of their atomic number (39).

Adsorption of the sample at the top of the cation-exchange resin column and its elution to yield the results shown in Figure 3 for this definitive experiment required the continuous attention of the researcher because no automatic control of effluent and sample collection had been provided. After a period of more than 30 hours had been spent by me to initiate and then attend to the experiment, fatigue had left me exhausted. My colleague, Glendenin, returning to the laboratory for another day of research, convinced me that I needed rest. It was late that afternoon that Glendenin awakened me to the news of our discovery of element 61.

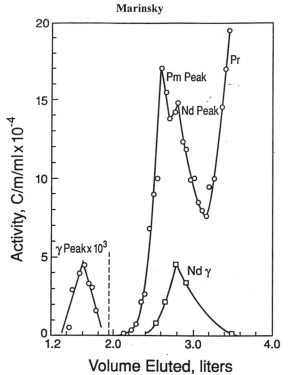

Figure 3 Rare-earth elution curve to establish element assignments; o=β elution curve with peaks for 57d Y^{91}, long-lived element 61, 11d Nd, and 13,8dPr. □=elution curve with peak for 11d Nd.

6.3. Discovery Confirmed

In the next few days the radioactivity associated with the element 61 (peak 2) and the neodymium (peak 3) was fully characterized. The radioactive isotope of element 61 that constituted peak 2 was a long-lived ($t_{1/2} \approx 4$ yrs), low energy (~0.2 MeV) beta emitter and corresponded to the rare earth fission product isotope discovered independently in earlier studies by Ballou (23) and by Goldschmidt and Morgan (42) who on the basis of its chemical behavior, had suggested that it might be an isotope of neodymium or praseodymium. Seiler and Winsberg (43), on the basis of their later examination of this radioisotope thought its assignment to element 61 or neodymium more likely. The radioactive neodymium isotope that constituted peak 3 was shown to be the 11 day gamma emitter (29) discovered earlier by Davies (25).

With the positive identification of long-lived element 61 and 11 day neodymium in the above definitive experiment (29,40,41) a more detailed study of these nuclides was conducted by me and Glendenin. Fission yields of the 11 day neodymium and the long-lived element 61 nuclides were determined to be nearly identical in value (29,40,41) and formed the basis for assignment of the same mass number to both nuclides. The parent-daughter relationship, implied by such an assignment, was confirmed experimentally. Our assignment of a mass number of 147 from consideration of the mass-fission yield curve, accessible as a major product of the nuclear chemistry programs led by Coryell and Sugarman (4), was corroborated in a mass spectrographic investigation by Lewis and Hayden (44) of an intensely radioactive sample of the highly purified element 61 nuclide that we prepared for this purpose.

The program of research was further extended, as well, to include study, with the ion-exchange procedure developed, of activities induced in neutron irradiated neodymium oxide (45). This search for additional element 61 and neodymium nuclides that had so far eluded identification resulted in the identification, with the ion-exchange approach, of a 47-hour activity as another isotope of element 61. The result of this experiment is summarized in Figure 4.

Once discovered as a product of neutron irradiation of stable neodymium, the 47 hour element 61 was looked for in fission. Irradiation of uranyl nitrate in the Clinton Laboratories' graphite reactor and eventual isolation of the element 61-neodymium fraction led to its observation (45). Study of the decay of the isolated element 61-neodymium fraction showed the presence of a 47 hour component, the 11 day neodymium, and negligible long-lived residual activity. Aluminum absorption studies of the beta radiation associated with the decay of the 47-hour activities produced in fission and by neutron irradiation of neodymium yielded identical results, thus establishing the existence of the 47 h element 61 as a fission product. Its fission yield was then determined. Reference to the mass-fission yield curve developed in the study of this aspect of the uranium fission process at Oak Ridge and Chicago (3) led to a probable mass assignment of 149 (45).

Finally, in a more detailed study of neodymium and element 61 activities induced by neutron irradiation of neodymium oxide two active species, a 1.7 hour and a 12 minute activity, were observed to be formed together with the 47 h $61^{(149)}$, the 11d Nd^{147} and the long-lived ($\sim$4y 61^{147}) element 61 nuclides (46). The 1.7 hour activity was identified as an isotope of neodymium by use of the ion-exchange resin adsorption-elution method. The fractionation that occurred between the 47 h $61^{(149)}$ and both the 11d Nd^{147} and 1.7 hour activities that eluted at the same rate in this experiment, identifying the 1.7 hour activity as an isotope of neodymium, is shown in Figure 5. On the basis of the fair agreement obtained between the observed and parent-daughter-based predictions of their activity ratios, the 1.7h Nd was assumed to be the parent of the 47h $61^{(149)}$ nuclide.

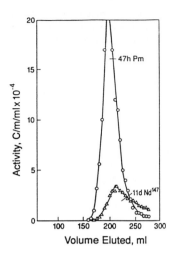

Figure 4 Elution curves of 47h element 61 and 11d Nd^{147} activities from neutron-irradiated Nd$_2$O$_3$. Elutriant pH = 2.82; activities are corrected to the end of the irradiation.

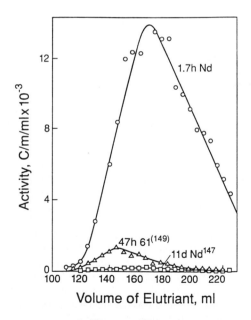

Figure 5 Elution curves of 1.7h Nd, 47h61$^{(149)}$, and 11d^{147}Nd.

Since at least three neodymium and three element 61 activities with masses of 147, 149, and 151, respectively, are to be expected on the basis of stable neodymium isotope occurrence, the 12m activity was tentatively assigned to $61^{(151)}$, and the $Nd^{(151)}$ parent was presumed to be too short lived for detection in our experiments.

While perception of this research by the scientists on the project was at a high level, it went unnoticed by the military. Two of the most exciting projects at the Clinton Laboratories in 1944 and 1945 had to be those dealing with the discovery of element 61 in fission and the successful development of a process for preparation of the intensely radioactive carrier-free ^{140}Ba-^{140}La source needed by Oppenheimer at Los Alamos to facilitate development of the trigger for the nuclear bomb under development. The more obvious military importance of the second project, however, led to the following response by the army to success in the two projects. Only one promotion to master sergeant was available to honor major contributions by military personnel in these research and development programs. It was, of course, awarded to the soldier (Strickland) who had contributed most importantly to the Ba^{140}-La^{140} isolation project. The scientific importance of the discovery of element 61 was disregarded by the army.

6.4. Announcing, Claiming and Naming Element 61

Disclosure of the research described above could not be made because of security restrictions specified by the United States Government. Even when Glendenin and I left Oak Ridge, Tennessee for Cambridge, Massachusetts in August 1946 to enter the doctoral program in nuclear chemistry that Professor Coryell had agreed to lead at the Massachusetts Institute of Technology, the research described above remained classified. Our frustration at being muzzled was removed in 1947 when declassification permitted announcement of our discovery of element 61 in the fission process at an American Chemical Society Meeting held in New York, September 1947. I presented our paper to over 1000 chemists crowded together in the Ballroom of the Pennsylvania Hotel.

Before this disclosure Glendenin and I had been examining the various claims to discovery of element 61 in nature and in cyclotron bombardments of praseodymium and neodymium. Urged on by Professor Coryell, we had decided, on the basis of uncertainties inherently present in these various claims, to present ourselves as the rightful discoverers of element 61 (47). When we suggested to Professor Coryell that he be listed as a codiscoverer of element 61 with us, he insisted that his contribution had not been sufficient to warrant this action. He pointed out, as well, that his name would distract attention from us because of his stature as a scientist. On the Manhattan Project his peers looked to him for technical and supportive guidance. He set an unmatched example of honesty and integrity for them with respect to credits for individual research efforts as well. It was our great luck to have had the opportunity to study and work with and to learn from Professor Coryell.

Our position was clearly enunciated in a press conference held after this session. This press conference was attended by Professor B.S. Hopkins who, together with Harris and Yntema, had in 1926 claimed the discovery of element 61 in nature and had proposed that the element be named "illinium" after the state and university in which their research had been performed (16). Also present at the press conference was Professor Quill (Ohio State University). He, together with Professor Pool, had suggested that element 61 be named "cyclonium", basing their claim of discovery on the production of radioactive isotopes of element 61 together with those of praseodymium and neodymium in cyclotron facilitated bombardment of stable praseodymium and neodymium targets (48,49).

Shortly after the American Chemical Society meeting the paper detailing our research of radioisotopes element 61 and neodymium at the Clinton Laboratories was published (40). Our proposal of the name promethium for element 61 appeared next in the Chemical and Engineering News (47). In it we asserted that we were the first to give positive evidence for the existence of an isotope of element 61. Strong arguments were presented to refute the earlier claims to such discovery and thereby to justify our objective, the acceptance of our right to name the element.

To negate the claim of Quill and Pool (48,49) a summary of the results of activity production in the cyclotron bombardments of neodymium and praseodymium targets (49-52) that provided the basis for their claim of discovery was presented (47). In the table provided for this purpose the activities observed, the assumed nuclear reactions and the proposed assignments of these activities by element and mass number were listed. Positive assignments for those activities accessible from our results were presented for comparison. We argued that the demonstratable uncertainty of their assignments, made without benefit of chemical separation, did not provide element identification, the prerequisite for acceptance of discovery claims (53). (See preceeding chapter for discussion of priority claims in the discovery of chemical elements).

In our consideration of an appropriate name, several of the candidates considered and then discarded were phoenicium (phoenix rising from the ashes of nuclear fission) and clintonium (after Clinton Laboratories). The final choice "promethium" was suggested by Charles Coryell's wife Grace Mary. The name refers to Prometheus, the Titan in Greek mythology, who stole fire from heaven for the use of mankind (54). It not only symbolizes the dramatic way in which the element is produced as a result of harnessing of the energy of nuclear fission, but also warns of the danger of punishment by the vulture of war.

On September 5, 1949 the name promethium was officially accepted by the International Union of Chemistry at its Amsterdam conference, and we were officially designated by the international community of scientists as its codiscoverers (55).

In 1948 G.W. Parker and P.M. Lantz isolated 4.5 mg of Pm^{147} (56). Its identity was established employing emission and X-ray spectroscopy (57). Since then kg quantities have been prepared from fuel reprocessing waste by F. Weigel and coworkers (58) who in 1963 prepared it in elemental form for the first time. About thirty promethium compounds were also prepared by Weigel's research group in a study of its chemistry in the 3+ state. The position of promethium between neodymium and samarium was demonstratable with these compounds.

Promethium-147 compounds are used for heat sources (SNAP generators), certain types of radiation sources and luminescent paints. Because of the low energy of its beta decay particle, the absence of gamma radiation and its fairly long half life Pm^{147} has also been used in pacemaker implants.

6.5. References

(1)   R.G. Hewlett and O. Anderson, Jr., "A History of the United States Atomic Energy Commission," University Park, Pennsylvania State University Press, 1962.

(2)   R.S. Mulliken, Radiochemical Studies: The Fission Products, National Nuclear Energy Series, Edited by C.D. Coryell and N. Sugarman, McGraw-Hill Book Company, Inc., New York-Toronto-London, Book 1, Introductory Note, xiii (1951).

(3)   "Radiochemical Studies: The Fission Products," National Nuclear Energy Series, Edited by C.D. Coryell and N. Sugarman, McGraw-Hill Book Company, Inc., New York-Toronto-London, Book 2, Introduction to Part V, Radioactivity of the Fission Products (1951).

(4)   J. Kleinberg, W.J. Argersinger, Jr., and E. Griswold, "Inorganic Chemistry," Chapter 26, D.C. Heath and Company, Boston, Ch 26, p 628 (1960).

(5)   W. Prandtl, Angew. Chem. 39, 896, 1331 (1926).

(6)   W. Prandtl, Ber., 60B, 621 (1927).

(7)   W. Prandtl and A. Grimm, Z. Anorg. Allgem. Chem. 136, 283 (1924).

(8)   J.G.F. Druce and F.H. Loring, Chem. News, 131, 273, 337 (1925).

(9)   C.A. von Welsbach, Chem. Ztg. 50, 990 (1926).

(10)   J.K. Marsh, J. Chem. Soc., 2387 (1929).

(11)   S. Takvorian, Compt. Rend., 192, 1220, 1972 (1931).

(12)   I. Noddack, Angew. Chem., 47, 301 (1934).

(13)   S. Takvorian, Ann. Chim., 20, 113 (1945).

(14)   M. Curie and S. Takvorian, Compt. Rend., 196, 923 (1933).

(15)   H.G. Moseley, J. Phil. Mag., 27, 703 (1914).

(16)   J.A. Harris, L.F. Yntema, and B.S. Hopkins, J. Am. Chem. Soc., 48, 1585 (1926).

(17)   L. Rolla and L. Fernandez, Gazz. chim. ital., 56, 435,688,862 (1926).

(18)   C.C. Kiess, Bur. Standards, Sci. Papers, 18, 201 (1922).

(19)   L.F. Yntema, J. Am. Chem. Soc., 46, 37 (1924).

(20)   J. Mattauch, Z. Phyzik., 91, 361 (1934).

(21)   B.S. Hopkins, J. Franklin Inst., 204, 1 (1927).

(22)   L. Rolla and L. Fernandez, Gazz. chim. ital., 57, 704 (1927).

(23)   N.E. Ballou, "Radiochemical Studies: The Fission Products," National Nuclear Energy Series, Edited by C.D. Coryell and N. Sugarman, Book 2, McGraw Hill Book Company, Inc., New York-Toronto-London, Paper 189, "Discovery of a long-lived rare earth activity", 1220-1224 (1951).

(24)   D.N. Hume and R.I. Martens, Clinton Laboratory Report CN-1311, June 6, 1944 (referred to in paper 191 of "Radiochemical Studies: the Fission Products," National Nuclear Energy Series) see reference 29.

(25)   T.H. Davies, Clinton Laboratory Report CN-1309, April 17, 1944 (referred to in paper 191 of "Radiochemical Studies: the Fission Products," National Nuclear Energy Series) see reference 29.

(26)   A.A. Noyes and W.C. Bray, "A System of Qualitative Analysis for the Rare Earth Elements," The Macmillan Company, New York, 1927.

(27)   N.E. Ballou, "Radiochemical Studies: The Fission Products," Edited by C.D. Coryell and N. Sugarman, Book 3, McGraw Hill Book Company, New York-Toronto-London, Paper 301, "Separation by alkaline fusion", 1722-1725 (1951).

(28)   F.T. Fitch, Ph.D. Thesis, Purdue University (1943).

(29)   J.A. Marinsky and L.E. Glendenin, "Radiochemical Studies: The Fission Products," Edited by C.D. Coryell and N. Sugarman, National Nuclear Energy Series, McGraw Hill Book Company, New York-Toronto-London, Book 2, Paper 191, "Identification and characteristics of 11d Nd^{147}", 1229-1242 (1951).

(30)   N.E. Ballou, ibid, Book 3, Paper 296, "Rapid Separation of Lanthanum and Yttrium Groups of Rare Earths", 1706-1709 (1951).

(31)   J.A. Marinsky, ibid, Book 3, Paper 297, "Behavior of Neodymium, Promethium, Samarium and Europium in Lanthanum-Yttrium Separations with Potassium Carbonate", 1710-1712 (1951).

(32)   J.A. Seiler, ibid, Book 3, Paper 298, "Separation of Lanthanum, Praseodymium and Neodymium by Potassium Carbonate", 1713-1715 (1951).

(33)   E.R. Purchase and E.J. Young, ibid, Book 3, Paper 299, "Composition of Precipitates of Lanthanum Formed with Sodium Carbonate", 1716-1718 (1951).

(34)   N.E: Ballou and J.A. Marinsky, ibid, Book 3, Paper 300, "Separation of Lanthanum from Praseodymium and Neodymium by Fusion with Sodium Nitrate", 1719-1721 (1951).

(35)   A.W. Adamson, A.R. Brosi, Q.V. Larson, L.S. Myers, Jr., J.A. Swartout, and N. Wells, Clinton Laboratory Report CN-3346, Dec 10, 1945.

(36)   A.W. Adamson, Clinton Laboratory Report CN-1859, April 19, 1944.

(37)   J. Schubert, Clinton Laboratory Report CN-1873, Jan. 1, 1945.

(38)   W.E. Cohn, E.R. Tompkins, J.X. Khym, G.W. Parker, S. Weiss, D.S. Ballantine, A.M. Ross and C.R. Vanneman, Clinton Laboratory Report CC-2827, June 1, 1945.

(39)   J.A. Marinsky and J.X. Khym, "Radiochemical Studies: The Fission Products," Edited by C.D. Coryell and N. Sugarman, National Nuclear Energy Series, McGraw Hill Book Company, New York-Toronto-London, Book 3, Paper 308, "Order of Elution Separation of Rare Earths from Amberlite Resins with Ammonium Citrate", 1747-1751 (1951).

(40)   J.A. Marinsky, L.E. Glendenin and C.D. Coryell, J. Am. Chem. Soc., 69, 2781 (1947).

(41)   J.A. Marinsky and L.E. Glendenin, "Radiochemical Studies: The Fission Products," Edited by C.D. Coryell and N. Sugarman, National Nuclear Energy Series, McGraw Hill Book Company, New York-Toronto-London, Book 2, Paper 192, "Identification and Characteristics of ~4y Pm^{147}", 1243-1253 (1951).

(42)   L. Goldschmidt and F. Morgan, Chalk River Laboratory Report MC-11, Aug. 14, 1943.

(43)   J.A. Seiler and L. Winsberg, "Radiochemical Studies: The Fission Products," Edited by C.D. Coryell and N. Sugarman, National Nuclear Energy Series, McGraw Hill Book Company, New York Toronto London, Book 2, Paper 190, "Further Study of the Long-Lived Rare Earth Activity in Fission", 1225-1228 (1951).

(44)   L. Lewis and R. Hayden, Clinton Laboratory Reports CP-3221, September, 1945 and CP-3295, Oct. 22, 1945.

(45)   J.A. Marinsky and L.E. Glendenin, "Radiochemical Studies: The Fission Products," Book 2, National Nuclear Energy Series, Edited by C.D. Coryell and N. Sugarman, McGraw-Hill Book Company, Inc., New York-Toronto-London, Book 2, Paper 193, "Discovery, Identification, and Characteristics of 47h Pm^{149}", 1254-1263 (1951).

(46)   Ibid, Book 2, Paper 194, "Study of Neodymium and Promethium Activities Induced by Neutron Irradiation of Neodymium", 1264-1272 (1951).

(47)   J.A. Marinsky and L.E. Glendenin, Chem. Eng. News, 26, 2346 (1948).

(48)   R.F. Gould, Chem. Eng. News, 25, 2555 (1947).

(49)   M.L. Pool and L.L. Quill, Phys. Rev., 53, 436 (1938).

(50)   H.B. Law, M.L. Pool, J.D. Kurbatov, and L.L. Quill, ibid. 59, 936 (1941).

(51)   J.D. Kurbatov, D.C. MacDonald, M.L. Pool and L.L. Quill, ibid. 61, 106 (1942).

(52)   J.D. Kurbatov and M.L. Pool, ibid. 63, 463 (1943).

(53)   F. Paneth, Nature, 159, 8 (1947).

(54)   G.T. Bulfinch, "The Age of Fable," Chap 2, pp 15-21; Bulfinch's Mythology, New York, The Modern Library (1855).

(55)   Committee on New Elements, IUPAC, Chem. Eng. News, 27, 2996 (1949).

(56)   G.W. Parker and P.M. Lantz, Oak Ridge National Laboratory Report, ORNL-75 (AECD-2160), June 18, 1948.

(57)   G.W. Parker and P.M. Lantz, J. Am. Chem. Soc., 72, 2834 (1950).

(58)   F. von Weigel, Chem.-Z., 102, No 10, 339 (1978).

EPISODES FROM THE HISTORY OF THE RARE EARTH ELEMENTS

PART II - APPLICATION

Carl Auer von Welsbach

CHAPTER 7

CARL AUER VON WELSBACH: A PIONEER IN THE INDUSTRIAL
APPLICATION OF RARE EARTHS

E. BAUMGARTNER
Johann Straußgasse 8/33
A-8010 Graz, Austria

7.1. Introduction

In this paper the contributions of Dr. Carl Auer von Welsbach to science and
technology will be divided into four sections. The order is not exactly determined by
the chronology of his life, since it makes more sense to deal with the different areas
that he studied in his scientific and industrial career. The sections are:

>   Auer - discoverer of chemical elements
>   Auer - inventor of the incandescent gas mantle
>   Auer - inventor of the osmium filament electric light bulb
>   Auer - discoverer of pyrophoric mischmetal alloys

As one can see, the main issue in Auer´s research was light, faithful to his life´s motto
"*Plus Lucis*" (More Light). In order to understand why the matter of light was so
important to him, one must remember how desperate the search for cheap, long lasting
sources of light was in Auer´s days. The industrial revolution was at full blast, and the
problem of illumination - before Auer´s invention only by the means of candles and
paraffin lamps - was urgent, especially in winter.

7.2. Biographical milestones

Carl Auer von Welsbach - Austrian chemist and entrepreneur, was born on September
1st, 1858 in Vienna as the 4th child of Alois Auer von Welsbach, director of the
Imperial and Royal Court Printery.

He received his primary education in Vienna, where young Carl developed his special
interests in subjects like chemistry, physics and geometry. He was impressed by the
cases of his father, himself a successful inventor. In 1877 he began his studies in

C. H. Evans (ed.), Episodes from the History of the Rare Earth Elements, 113–129.
© 1996 *Kluwer Academic Publishers. Printed in the Netherlands.*

chemistry at the university of Vienna. This part of Auer's life will be discussed in more detail in the next section.

Auer von Welsbach as a schoolboy

Within a 20 year period of his life, Auer made all his scientific and industrial inventions. During that time Auer moved to Carinthia, the most southern province of Austria. There he found what was important to him: an undisturbed atmosphere to conduct his research, an area of traditional industry based on iron ore, as well as widespread woods for his beloved harmony with nature including hunting and fishing.

He married Maria Nimpfer in 1899 and had 4 children. In honour of his outstanding work he was granted a baronetcy by the Austrian emperor in 1901. He received all thinkable honours of his time, among those honorary doctorates of several European universities and in 1920 the famous Siemens ring. He is the only chemist who was ever pictured on a bank note, the Austrian 20 Schilling note issued after World War II, as well as on an Austrian stamp. He died in his castle Rastenfeld near Treibach-Althofen in Carinthia, on August 4th, 1929 at the age of 71 after a life devoted to the research and industrial development of the rare earths. Auer´s outstanding scientific and industrial career has been described in the literature by several authors (1-4).

Carl Auer von Welsbach's picture on the Austrian 20 Schilling note

7.3. Discoverer of elements

Before discussing Auer´s part in solving the mystery of the rare earths, which puzzled chemists for over 150 years, the early studies of the rare earths will be summarized; (see chapters 1-5 for details).

It all started in 1787, when the Swedish army lieutenant Axel Arrhenius,who also happened to be a passionate and gifted amateur mineralogist, discovered a yet unknown black mineral in a quarry near the Swedish town of Ytterby.

The Finnish chemist Johan Gadolin, a professor at the University of Åbo, today´s Turku, became interested in the material. In 1794 he claimed that he had found a new "earth" in Arrhenius´s "black stone". The mineral was originally named "ytterbite", but was changed to gadolinite after this famous chemist. An "earth" was considered a homogeneous element then, at a time when neither atomic structure nor the periodic system of the elements were known. Today we know that the "earths" of the 18th and 19th century were really metal oxides; or rather mixtures of different metal oxides. Gadolin´s earth was named yttria after Ytterby, where it was first found.

In 1804 the German investigator Klaproth and the Swedish chemists Berzelius and Hisinger independently found another new earth with properties similar to yttria in a mineral which had been found in a mine close to the Swedish town of Bastnäs. Berzelius and Hisinger named their finding "ceria" after a planetoid that had been recently discovered.

Yttria and ceria were the two "roots" for the discovery of rare earths. Even today we divide the rare earth elements into two groups - the yttrics and the cerics. For the next century chemists studied the two earths and discovered that both consisted of several individual elements. New fractions were found, mainly through fractional precipitation, and new elements were proposed which in turn proved to be mixtures. Chemical properties were disputed and discoveries by one scientist were doubted by others. There were also many false conclusions.Several new elements were proudly pronounced, a paper written, and later discovered that this element did not exist. This part of the history of rare earths was not as orderly as it seems to us today, when we look at the clear and logical periodic table of the elements. Considering the little means and the disputed definitions -neither "element" nor "atomic weight" were defined undisputedly- that chemists were confronted at that time. Their search for new elements required not only time and effort, but also an enormous diligence, creative thinking, and the ability always to doubt one´s own findings.

Ceria had been divided into 3 fractions - cerium, lanthanum ("escaped notice") and didymium ("twinship") and yttria, terbia and erbia had been found in the yttrics by the time Auer was born in 1858. He was still in his childhood when Bunsen and Kirchhoff invented the spectral analysis, which was crucial to determine the purity of a newly discovered "element".

Robert Wilhelm Bunsen, Auer von Welsbach's mentor

When Auer was 20, he began to study chemistry in Vienna with Prof. Adolf Lieben. He recommended Auer to his teacher Prof. Robert Bunsen, Germany`s outstanding scientist at that time, best known for the Bunsen burner and his work on spectral analysis. In 1880 when Auer went to Heidelberg, ytterbium, thulium, holmium and scandium had been discovered from the yttric group. Samarium had also been divided from didymium in the ceric group. As a student of Bunsen, Auer came in contact with the two ideas that would determine his scientific career: the fascinating world of rare earths and Bunsen's idea of enhancing the light of a burner's flame.

Auer and Bunsen remained close until Bunsen's death in 1899. Auer never forgot Bunsen's impact on his life and as his mentor he should always be the shining example of a scientist as he wrote in a letter in memory of his teacher (5). Auer purchased Bunsen's library from his estate, but like his teacher he derived little pleasure from reading; and so many of Bunsen's books remained unopened, as he had received them.

Auer, who had never submitted a dissertation, got his PhD on May 2nd, 1882 in Heidelberg (6). Bunsen was convinced of the scientific abilities of his gifted student and thought it did not require written proof (3). Bunsen had the opinion that a good knowledge of physics was the necessary foundation for a successful career in chemistry.

Auer returned to Vienna in 1882 to work with Prof. Lieben as an unpaid aid. His father had left him a large inheritance, which allowed him to pursue his research interest in the field of rare earths. In 1883 he published his first paper, dealing with the separation of gadolinite (7,8). Instead of following the traditional method of separating the rare earth minerals by fractional precipitation, he developed a new technique, the fractional crystallisation of the ammonium double nitrates in concentrated nitric acid. Compounds of individual rare earth elements exhibited a difference in crystallisation; thus he was able to separate them.

He suspected didymium was not homogeneous since samples prepared from different minerals had a different absorption spectrum and different atomic weights. With his new method, Auer found a green fraction of didymium which he called praseodidymium ("green twin") and a pink fraction which he called neodidymium ("new twin"). The detection of these new elements was published in 1885 (9). In 1903 he published a 2nd, refined paper on this separation (10). He had calculated the atomic weight of praseodymium to be 140.57 and that of neodymium to be 144.54. For comparison, the atomic weight of praseodymium is documented today at 140.9077, while that of neodymium is 144.24.

But those two elements would not be the only ones Auer discovered. Already a successful businessman he was still interested in research. He turned to the separation of the yttric group. From the results of the spectral analysis of ytterbium, by 1905 he already believed it was not homogeneous (11). Using the fractional crystallisation of ammonium double oxalates, he was able to separate the impure ytterbium into two elemental compounds. In 1906 he succeeded in preparing them and gave them mythological names: aldebaranium (element 70) and cassiopeium (element 71) (12). Although he had already named the new elements it took him a long time to prepare the final publication.This hesitation turned out to be a big mistake.

The French scientist, Georges Urbain, had also suspected ytterbium was not homogeneous. He used the method of the fractional crystallisation of rare earth nitrates in the presence of bismuth nitrate. Urbain also found two fractions within the impure ytterbium. He named them neoytterbium (element 70) and lutetium after the ancient name of Paris (element 71) (13). He gave a speech on his findings on November 4th, 1907 in Paris.

Forty-four days later, on December 19th, 1907, Auer made his final presentation (14). He characterized the new elements based on accurate information concerning the spectral lines and presented calculations of the atomic weight: 172.9 for aldebaranium (today: 173.04) and 174.23 for cassiopeium (today: 174.967).

A priority dispute started. Both scientists claimed the glory of the discovery and the rights of naming the new elements. The increasing political antagonism between France and Austria/Germany prevented a compromise. In 1909 the International Commission of Atomic Weights gave Urbain, who was a member of this commission, the priority. The commission's decision had ignored Auer's previous papers from 1905 and 1906 and had only regarded his publications from late 1907. Urbain had indeed been faster. Auer was very disappointed and explained in an additional paper on the "Separation of Ytterbium" in 1909 that his original results had been far more accurate and the compounds had been purer (15,16).

The German Atomic Weights Commission refused to accept the decision of the International Commission and decided in 1923 to use the names ytterbium and cassiopeium (17), which continued until World War II. Today element 71 is attributed to both Auer and Urbain, but the name lutetium has remained.

This discovery was one of the last chapters in the discovery of rare earth elements. However, since there was no theoretical explanation of the atomic structure and the similarity of rare earths, and therefore no limit to the number of rare earth elements, chemists kept searching for more rare earth elements. The rivalry between Auer and Urbain continued. Both claimed to have found more elements. Auer, for example, thought that thulium could be separated in three individual elements (18,19). Urbain reported from his own lutetium in 1911 a new element 72, which he named celtium. After more careful investigations it turned out that it was high purity lutetium, identical with Auer's cassiopeium.

Auer´s last publication also dealt with the search for new elements; he tried to fill the blank space of element 61 and reported in 1926 that he could not find an element between neodymium and samarium (20). From today´s point of view we know that element 61, promethium, is an unstable radioactive element that was not isolated until 1947 (see Chapter 6).

Georges Urbain, Auer von Welsbach's nemesis

Front page of the Viennese newspaper "Extrablatt" from February 1, 1886. The picture shows Carl Auer von Welsbach in his laboratory at the Institute of Chemistry of the University of Vienna as the inventor of the incandescent light. In the upper left corner an example of the structure of the new light source can be seen. This historic laboratory has been reconstructed in the technical museum of Vienna.

7.4. The incandescent mantle

In the same year that Auer succeeded in the separation of didymium, Auer made an invention that revolutionised illumination and led to the first industrial application for rare earths.This invention also brought new perspectives to rare earth research, which had only been conducted for the sake of science up to then. Now, with a practical use for them, further knowledge about the 4f-elements not only meant scientific fame but also worldly riches. And a rich man Auer became; but first things first.

In his days in Heidelberg, Auer had learned from Bunsen the incandescent principle: that the flame of a gas burner, later on named the Bunsen burner, could be enhanced through the influence of metal oxides. In 1885 the first two patents for incandescent gas lamps were granted. Auer saturated a cotton fabric with rare earth nitrates and then burned the fabric. A net-like structure of rare earth oxides remained, which was put over the burner flame to increase the emission of light (21).

A mixture of lanthanum oxide and zirconium oxide proved to amplify the light best. Auer called this mixture "actinophor" and patented it in 1886. Unlike other scientists, Auer was an excellent businessman. He patented his invention in various countries, produced the "fluid" himself and gave out licences to companies in other countries. In 1887 he bought a bankrupt pharmaceutical factory in Atzgersdorf near Vienna to manufacture lanthanum nitrate by the means of fractional crystallisation on a large scale. He then sent the solution to his licensees, who produced the mantles, but kept the composition of the lighting fluid secret.

But problems soon followed the first success: the light was too green, the mantle did not last, and it was too expensive; competition from the electric light became stronger. The licencees threatened to sue Auer, who had to shut down his plant in Atzgersdorf two years after it had opened. Nevertheless he did not let this failure discourage him. In his own private laboratory, he tried new mixtures of oxides and searched for new raw materials.

A cheap new source of raw material was found overseas - monazite sand, which occured abundantly in the United States and Brazil . It consisted of rare earth compounds (cerium, lanthanum, neodymium and praseodymium) and a considerable amount of thorium. Auer soon found that this new by-product of fractional crystallisation enhanced the flame of the burner to a soft glowing light and lasted longer than his old "actinophor". He also found that the light of the thorium mantle was better if the thorium was less pure. It did not take long to find that the main impurity was cerium; and he found the "perfect" composition for his new and improved incandescent light - 99% thorium and 1% cerium. When this new illumination was introduced to the public on the 4th of November, 1891 outside the Opern Cafe in Vienna, it was the hour of birth for the rare earth industry. Auer had finally found the first industrial application for the still mysterious rare earths.

The new mixture was a success all over the world. The plant in Atzgersdorf reopened, and 90,000 lamps were sold in the first 9 months of large-scale production. In 1913 worldwide production was 300 million.

The Auer light was mainly used indoors, the biggest buyer being railway companies, who continued to use the incandescent light long after electric illumination was used elsewhere, since it was cheaper. But it also worked outdoors, and a couple of cities followed Bombay's example - this Indian city was the first to introduce public lighting with Auer's lamp. Today we cannot imagine what Auer's invention meant to the world of candles and paraffin then. To some people the soft light of the incandescent mantle, which lasted long, didn't smell, and was neither expensive nor hazardous, seemed like a miracle.

7.5. The metal filament lamp

In 1893 Auer bought an estate with a castle in the South of Austria, Schloß Rastenfeld, which he made his home and place of retreat. There he could pursue his two favourite activities - hunting in the woods around his castle and research in his new private laboratory.

In 1898 Auer offered to buy an ironwork in the nearby village of Treibach that had been closed 11 years earlier. There he founded "Dr. Carl Auer von Welsbach's Werk Treibach" (22). The iron work had electricity, needed for his research in reducing rare earth salts to metals as well as for developing his idea of a metal filament electric light bulb.

Edison's invention of the carbon filament lamp in 1879 was no serious threat to Auer's incandescent light, because of a bad economy. Auer turned to the idea of a metal filament electric lamp. He used what he thought was the metal with the highest melting point - osmium. The main problem Auer faced with this project was that metallic osmium is very hard and brittle, which made it difficult to shape into thin filaments. No procedure was yet known to shape this metal.

The first method he tried to obtain osmium filaments was the alloying method: fine platinum wires were put into a tube filled with a reducing agent and gaseous osmium oxide (OsO_4). When the wires were heated with electric current, metallic osmium was refined. This process was repeated until the filament was thick enough. The drawback of this method was that the filaments did not all have the same diameter and were not flexible enough.

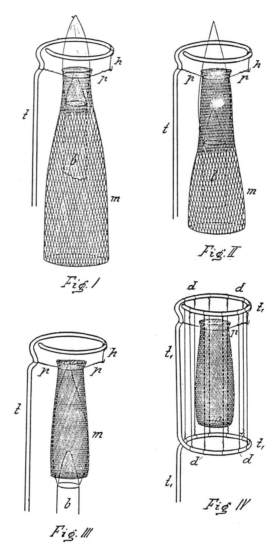

Examples of incandescent mantles designed by Auer von Welsbach consisting of a cotton fabric suspended from a metal wire. The figures were shown in the journal "Der Gastechniker" (The gas technician), in 1887.

Auer developed a new and improved procedure in which osmium chloride was thermally reduced to amorphous osmium. A paste was made out of this amorphous osmium with collodion or sugar. This paste was pressed through fine diamond tubes, cut, shaped and made red-hot in a vacuum, to remove the carbon. The result was an osmium filament that could be put into a bulb, for which Auer was granted a patent.

Auer´s patent in 1893 was the "birth certificate" of the metal filament lamp. The osmium filament electric bulbs were fabricated in Atzgersdorf. The lamps were not sold, but were leased from Auer´s company. Thus he overcame the problem of the short life of the filaments. In comparison with the carbon filament lamp the osmium lamp required only 1.5 watts for one candle power, versus 3.5 watts for the former.

The osmium filament lamp was introduced to the public at the world exposition in Paris in 1900, which was also illuminated with Auer´s incandescent light.

Auer bought all the osmium supply he could get, but the osmium lamp was not a long lasting success. He had overlooked the two metals with an even higher melting point, tantalum and tungsten. Other inventors used Auer´s ideas with these metals. Siemens and Halske were the first to introduce a tantalum filament lamp. The final success however belonged to the tungsten filament lamp.

But even if Auer was denied the economic success of his invention, one cannot deny him the honour of being the first to develop the metal filament lamp and a procedure to shape hard and brittle materials. The fact that he tried to improve electric illumination, although this meant competition for his gas mantle business, showed that even more than a businessman, Auer was a scientist and inventor; a man whose quest to improve living conditions was the driving force in his thinking.

7.6. Mischmetal and pyrophoric alloy

The roots to this invention again are to be found during Auer´s studies with Bunsen. While in Heidelberg he learned about fused salt electrolysis for the production of rare earth metals. It attracted Auer´s attention that the metal at the rim of the crucible, where it was alloyed with iron from the electrodes, sent out sparks when filed. Neither pure iron nor pure rare earth metals showed this property.

Auer remembered this observation when he thought of the large quantities of rare earths that were left over from the production of cerium and thorium for the incandescent mantle fluid. He looked for an industrial use of these by-products. Auer was fascinated by the idea of a cheap and easy way for the ignition of incandescent light.

Auer´s imaginative mind found one solution to both problems. In 1903 the "pyrophoric metal alloys" were patented. Auer had coined the term "pyrophoric" to describe materials that sent out sparks with little mechanical work applied. The sparks could be used to ignite inflammable materials.The patent was strenously litigated, but this time Auer won the priority dispute. Thus, in addition to having found the first industrial application of rare earths, Auer had also found the first metallurgical application of rare earths.

Intensive research was performed in Treibach, because although the new alloy was patented in 1903, there was still a long way to go before beginning of industrial production. The best pyrophoric property was found in an alloy of 70% mixed rare earth metals and 30% iron (23). The material however was very brittle and had pores that needed to be eliminated.

In order to minimize the losses caused by this long period of research, Auer separated his research and development facility in Treibach from his real estate and founded the "Treibacher Chemischen Werke", a private limited liability company in 1907 (22).

In 1908 he and his general manager, Dr. Fattinger, succeeded in eliminating the pores in the mischmetal, caused in part by phosphorous, a natural constituent of Auer´s raw material monazite, and in part by the content of oxychloride resulting from the drying process of the mixed rare earth chloride. With this pore-free mischmetal, Auer was able to manufacture the first corrosion resistant lighter flints. The invention was a huge success and applications such as replacement of the starter motor for automobiles were considered for "Auermetal". In 1908, 800 kg of mischmetal were processed into 4 million lighter flints. The first pocket lighters were also produced in Treibach. The lighter flint was the most important application of rare earths until the 1930s and required about 1500 tons of rare earth chlorides per year.

At this point it should be mentioned, that Auer gave also an important impulse for research into radioactive phenomena in their early days. In his factory in Atzgersdorf the separation of radium from uranium - pitchblende found in Joachimsthal was carried out from 1904 on. This factory had the largest separation capacity at that time. Back then Joachimsthal was part of the Austrian - Hungarian Empire; today it is the Czech Republic. It is reported, that more than 5 grammes of radium were produced as radium-barium chloride, the largest amount at the time (3). Out of this sample the first atomic weight determination of purified radium chloride was performed. In 1908 Rutherford was given 400 mg of radium chloride, with which he performed his famous experiments leading to Rutherford`s atomic model. Radium was sold by the Austrian Academy of Science to research institutes in many countries, including the United States, United Kingdom, Russia, and Canada. With the money received, the Radium Institute of the Austrian Academy of Science in Vienna was founded. Next to the equivalent institute in Paris, this was the most famous radium institute in the world. Looking back, one can say that the separation technique invented by Auer laid the foundation for important scientific discoveries in such fields as those known today as radiochemistry and nuclear physics.

Auer von Welsbach, towards the end of his life

Only three applications of rare earths materialised throughout the 1930s, and two of them stemmed from Auer von Welsbach´s practical, creative mind. He was not the typical scientist of today, he enjoyed neither reading nor publishing and always looked for practical aspects of his ideas. He was definitely not an *"l´art pour l´art"* chemist. In addition to his inventions and the founding of the "Treibacher Chemischen Werke", he is also remembered because of his social attitude in attempting to improve the lives of his workers, and his charitable activities for their families and children. In many ways he gave more light to the world.

As mentioned, Auer von Welsbach died on August 4th, 1929. His life was devoted to natural science, thus obeying the law of nature. In memory of this great Austrian chemist and inventor, the Austrian Society of Chemistry awards the "Auer von Welsbach Medal" to persons making important contributions to the chemical industry.

7.7. References

1. D´Ans, J., *Berichte der dt. chemischen Gesellschaft*, 64 (1931) Heft 5 Abt.A.

2. Peters, K., *Blätter für Technikgeschichte*, 20 (1958).

3. Gutman, V., *Journal of Chemical Education*, 47 (1970) 209.

4. Szabadvary, F., *Handbook on the Physics and Chemistry of Rare Earths*, 11 (1988) 33.

5. Auer von Welsbach, C., *Brief an Prof. Dr. Paul Askenary*, (1911)

6. *Letter from the library of the University of Heidelberg to the library of the University of Vienna* (1958).

7. Auer von Welsbach, C., *Monatsh. Chem.*, 4 (1883) 630.

8. Auer von Welsbach, C., *Monatsh. Chem.*, 5 (1884) 1.

9. Auer von Welsbach, C., *Monatsh. Chem.*, 6 (1885) 477.

10. Auer von Welsbach, C., *Sitz.-ber. Akad. Wiss.*, 112 II a (1903) 1037.

11. Auer von Welsbach, C., *Akademischer Anzeiger*, 42 (1905) 122.

12. Auer von Welsbach, C., *Monatsh. Chem.*, 27 (1906) 935.

13. Urbain, G., *Compt. rend. Acad. Sciences*, 145 (1907) 759

14. Auer von Welsbach, C., *Monatsh. Chem.*, 29 (1908) 181.

15. Auer von Welsbach, C., *Monatsh. Chem.*, 30 (1909) 695.

16. Auer von Welsbach, C., *Zeitschr. Anorg. Allgem. Chem.*,67 (1910) 149.

17. *Ber. dt. Chem. Ges.*, 57 A (1924) 5.

18. Auer von Welsbach, C., *Monatsh. Chem.*, 32 (1911) 373.

19. Auer von Welsbach, C., *Zeitschr. Anorg. Allgem. Chem.*, 7 (1911) 439.

20. Auer von Welsbach, C., *Chem. Ztg.*, 118 (1926) 990.

21. Auer von Welsbach, C., *Wochenschrift des Niederösterreichischen Gewerbevereins*, (1886).

22. Smetana, O., Dauschan, W., *Treibacher Chemische Werke, Entstehung und Entwicklung*, (1980) 20.

23. Fattinger, F., *Chem.Ztg.*, 53 (1910) 469.

CHAPTER 8

THE HISTORY OF CHINA'S RARE EARTH INDUSTRY

WANG MINGGIN
 and
DOU XUEHONG
China Rare Earth Information Center
Baotou Research Institute of Rare Earth
Inner Mongolia

8.1. Introduction

With the development of modern technology, the world's rare earth industry has been expanding rapidly for the past thirty years. China, thanks to its abundant rare earth resources, is endowed with favourable conditions for developing its rare earth industry, to which great importance has also been attached, especially in the last 10 years. Influenced by the implementation of reform and open policy, and inspired by the great demand at home and abroad, production of rare earths has been sharply increased and their application has been extended to various fields. Today, China has become not only the major supplier of rare earth raw matrials, but also the major producer, consumer and exporter of rare earth products, taking an important place in the world market.

8.2. Resources

8.2.1. SPECIALITIES

China's rare earth resources are characterized by their large reserves, multiple types, wide distribution, variety, high content of valuable elements and high ratio of intergrowth minerals. Analysis reveals that the recoverable reserves amount to 36 million tons of rare earth ores (REO), making up nearly 75% of the world's total reserves of 48 million tons. The type, distribution and dates of discovery of China's major rare earth deposits are listed in Table 1.

C. H. Evans (ed.), Episodes from the History of the Rare Earth Elements, 131–147.
© 1996 *Kluwer Academic Publishers. Printed in the Netherlands.*

TABLE 1 Major Rare Earth Deposits in China

TYPE	AVERAGE REO CONTENT	SOURCES	DATE OF DISCOVERY
Mixed RE ore	5-7%	Baiyunebo, Inner Mongolia	1935
Ion-adsorption	0.08-0.15%	Jiangxi, Guangdong, etc	1969
Bastnaesite	0.5-5%	Shangdong, Sichuan	1960s
Monazite	500-1000g/m^3	Guangdong, Hainan, etc	early 1950s
Xenotime	80-200g/m^3	Guangdong, Hainan, etc.	early 1950s
RE-containing Collophanite	0.03-0.16%	Guizhou	early 1980s

REO = Rare Earth Ore
Source: China Rare Earth Factual Database

8.2.2. MAIN INDUSTRIAL DEPOSITS

The supply of raw material to China's rare earth industry relies mainly on rare earth deposits of two large types. The deposit in Baiyunebo, Inner Mongolia, contains mixed light rare earths while the ion adsorption rare earth deposits found in such places as Jiangxi and Guandong are rich in medium and heavy rare earths. A small proportion of the supply also comes from bastnaesite, monazite and xenotime.

8.2.2.1. *Baiyunebo Mine*
Baiyunebo ore mine, 135 km north of Baotou, is the largest rare earth deposit in the world. The rare earths form intergrowth minerals with iron, niobium, etc. Their recoverable reserves amount to 35.05 million tons, and the total size of the deposit may exceed 100 million tons. The mine was discovered as an iron deposit in 1927 by Professor Ding Daoheng, a well-known geologist. In 1935 Professor He Zoulin, a Chinese mineralogist, confirmed that the Baiyunebo ore contains two types of minerals: bastnaesite and monazite. A detailed geological survey of the whole mine was made during the 1950s, after which the mine was built and operated as the iron ore base of the Baotou Iron and Steel Company. The recovery of rare earths during the process of making iron and steel, however, began at the end of the 1950s.

Baiyunebo ore is a mixed complex rare earth ore -- the intergrowth of bastnaesite and monazite. The ratio of the former to the latter in the ore body is approximately 6-7:4-3. In comparison with bastnaesite from the Mountain Pass, the bastnaesite mentioned above contains higher samarium and europium, and the monazite is low in thorium and is free from uranium. Such rare earth minerals are closely associated with iron-, niobium-, and titanium- bearing minerals and are embedded with finely divided particles, thus making them very difficult to separate. However, the rare earths needed for industry are obtained as a by-product of the extraction of iron from its ore, thus lowering the operating cost.

8.2.2.2. *Ion Adsorption Type Deposits*
The rare earth elements in this unique ion adsorption type of deposit exist as ions adsorbed onto clay. They are generally rich in medium and heavy lanthanides. The exploitable reserves throughout the country have been verified as more than 500,000 tons, and the potential reserves have been estimated at 10 million tons. The rare earth content of the deposit ranges generally from 0.08 to 0.15%. All of the ore bodies are exposed near the earth's surface, which makes them easy to extract. The deposit has now become the major souce of medium and heavy rare earths in China.

The occurrence of the ore was first discovered in the granite crust weathering in Longnan county, Jiangxi Province, in 1969, and was afterwards found in the whole area of south Jiangxi, as well as Guandong, Guangxi, Hunan, and Fujian. Exploitation of the ore has been carried out to varying degrees.

8.2.2.3. *Bastnaesite*
In China, simple bastnaesite deposits were discovered in Weishan county, Shandong Province, and Mianning County, Sichuan Province in the 1960s and 1980s respectively. In the Weishan mine, mineral veins rich in rare earths have been found and proved to be of high purity and coarse simple mineral bastnaesite. The Mianning ore body mainly consists of bastnaesite with small amounts of parisite, chevkinite and xenotime. Minerals containing lead, molybdenum and bismuth have also been found in association with rare earth minerals.

Both the Weishan and Mianning minerals favour the separation of rare earths, and their separability is better than that of Baiyunebo mixed minerals, thus giving high recovery and low cost.

8.2.2.4. *Monazite and Xenotime*
Monazite and xenotime, mainly dispersed over Guangdong, Hainan, Guangxi and Hunan provinces, are mostly obtained as a by-product in processing ilmenite-, zircon- and fergusonite- bearing coastal heavy placer or river placer deposits. China's industrial reserves of rare earths contained in monazite amount to 480,000 tons. In the early 1960s the heavy placer deposits were exploited and utilized, and monazite at once became one of the principal sources of China's rare earth supply. Presently the extraction of monazite from coastal placer is carried out in Guandong and Hainan provinces. Of the two minerals, the Nanshanhai mine (Guangdong) has a higher grade of rare earths; it contains $1084g/m^3$ monazite and $200g/m^3$ xenotime. Though the technology for processing these minerals is practical and was earlier adopted in industry, they have been replaced gradually by other rare earth ores because of their

higher content of thorium and uranium, which demands strict safety protection and waste slag disposal.

8.2.2.5. *Rare Earth containing Collophanite*

In the early 1980s large amounts of rare earths, rich in the yttrium subgroup held in collophanite, were found in Zhijin county, Guizhou Province. The crude ore contains 0.03-0.16% REO. The total rare earth reserves have been estimated at 700,000 tons. Since the rare earths disperse in the mother ore in the form of isomorphs and cannot be concentrated by means of physical separation, they have not been exploited and utilized so far.

8.3. Production

Rare earth production in China began in the middle 1950s. In those days only monazite was processed to produce rare earth metals for lighter flints and thorium nitrate for gas mantles (see Chapter 7). Since the 1960s great importance has been attached to a more comprehensive utilization of the Baiyunebo deposit. For this, technical personnel associated with rare earths throughout the country have been deployed to undertake research into comprehensive recovery of rare earths, as well as their popularisation and application, thus promoting all-round development of the rare earth industry. With the world's increasing demand for rare earths, development has occurred ever more rapidly. During the years 1978-1989, the increase in production averaged 40% per annum (see Table 2), making China one of the largest producers of rare earths in the world.

TABLE 2 China's Rare Earth Production 1978-1990 (t, REO)

YEAR	OUTPUT	YEAR	OUTPUT
1978	1000	1985	9500
1979	1500	1986	11860
1980	2524	1987	15100
1981	2778	1988	18660
1982	2969	1989	19670
1983	3900	1990	14964
1984	6000		

8.3.1. RAW MATERIAL AND PROCESSING TECHNOLOGY

Over the past 30 years a series of rare earth production processes with distinctive national features has been developed, taking into account the characteristics of national resources and technical and economic conditions. Baiyunebo mixed ore and ion adsorption ore are the chief sources of China's rare earth raw material. As they are peculiar ore types, their method of processing is quite different from the conventional one. Minerals such as bastnaesite, monazite and xenotime are also exploited and utilized, but they occupy only a small fraction of the whole raw rare earth material and their method of processing is basically similar to that adopted abroad.

8.3.1.1. *Extraction of Rare Earths from Baiyunebo Ore*

The Baiyunebo ore comprises as many as 14 rare earth minerals growing together in a complicated fashion. Among these minerals bastnaesite and monazite predominate. The specific gravity of these minerals approximates that of iron minerals and barite; their magnetism approximates that of haematite, soda pyroxene and riebeckite, all being minerals of weak magnetism; their floatability is close to that of calcite, fluorite, barite and apatite; and their embedded particle size generally ranges from 10-70µm (particle sizes smaller than 40 µm account for 77%), so the enrichment of rare earths by traditional methods of separation is difficult. Currently, two processing routes for extracting rare earths from Baiyunebo ore are in operation.

8.3.1.1.1. *Blast furnace smelting--slag rich in rare earths--rare earth master alloy* In 1956, the research team led by Professor Zou Yuanxi, a well-known metallurgist of the Shanghai Institute of Metallurgy, Academia Sinica, creatively succeeded in developing a process for making RE-Si-Fe alloy, in which rare earths were recovered from blast furnace slag containing 4-6% REO by using ferrosilicon as a reductant. At the end of the 1950s the process was set in operation in No. 1 Rare Earth plant of BISC, thereby initiating the history of recovering rare earths from the Baiyunebo deposit and making RE-Si-Fe alloy in China.

In 1965, a number of researchers from the Baotou Research Institute of Rare Earth (formerly Baotou Research Institute of Metallurgy) successfully worked out a new process for increasing the rare earth content of the slag mentioned above. In this process, medium- and low-grade iron ore (containing 25-35% TFe and 7-9% REO) or low-grade rare earth concentrate pellets (containing 15-30% TFe and 12-15% REO) are directly fed into a special blast furnace to make a slag rich in rare earths (containing 10-16% REO) by removing iron and other unwanted elements. The slag is then processed silico-thermo-electrically to produce RE-Si-Fe alloy. After a series of pilot-plant tests conducted in the Shanghai Ferroalloy Plant in 1966, regular commercial production of the slag was carried out in Baotou Rare Earth Ferroalloy Plant (formerly Baotou Dongfeng Iron and Steel Plant) in 1969. The slag obtained contains over 100% more REO than the common blast furnace slag of BISC. This increase in REO greatly reduced the consumption of energy and raw materials, and increased the recovery of rare earths. This, in turn, lowered the price of rare earths and created favourable conditions for promoting their application in iron and steel manufacturing.
This process has been used so far to produce rare earth master alloy in China.

8.3.1.1.2. *Ore-dressing --- rare earth concentrate -- rare earth product* Twenty years have been spent in developing an ore dressing technology applicable to low- and medium-grade Baiyunebo oxidized ore. The research was carried out cooperatively by researchers from the General Research Institute of Mining and Metallurgy, Beijing;

Baotou Research Institute of Rare Earth; Guangzhou Research Institute of Non-ferrous Metals and some other institutions. In 1980 a breakthrough was eventually made, and a process for obtaining 60% REO came into being. The process has been put into operation in the Ore Dressing Plant of BISC.

Of the two kinds of concentrate produced according to the process, the one which is low in rare earths can be fed into an electrical submerged-arc furnace after pelletisation. The slag thus obtained is used to prepare rare earth master alloy. The phosphorous-containing pig iron obtained during the course of diferrisation is in accord with the standard YB 524-65 ferro-phosphorus and contains 0.2-0.6%Nb. It can be marketed as a by-product. In the late 1980s the Nanjing Reserch Institute of Metallurgy succeeded in preparing rare earth master alloy by directly feeding the low-grade rare earth concentrate into an electrical arc furnace to produce rare earth master alloy, thereby simplifying the process, reducing energy consumption and raising recovery.

A variety of rare earth master alloys have been produced in China. Most of these alloys, except RE-Si-Fe alloy, are prepared by melting RE-Si-Fe and other metal materials.

High-grade rare earth concentrates can be decomposed by the following methods: (1) concentrated sulphuric acid roasting, (2) sodium hydroxide decomposition, (3) sodium carbonate roasting, and (4) pyro-chlorination. The first two methods are currently used in production.

On the basis of the traditional method and through its constant improvement, the concentrated sulphuric acid roasting process was eventually developed by the General Research Institute of Non-ferrous Metals, Beijing in 1978-1979 and was put into operation in the early 1980s.

The advantages of the method are (1) low production cost, (2) simple operation, and (3) high adaptability for various raw materials; its disadvantages are (1) large amount of acidic waste gases and slag, and (2) serious corrosion of equipment.

The sodium hydroxide decomposition method developed in 1979 by the Baotou Research Institute of Rare Earth has an advantage over the sulphuric acid method in that it produces less waste gases and residue and provides better labour conditions for workers. The drawbacks of the process are (1) it is higher in production costs and (2) it produces larger amounts of waste water. To overcome these drawbacks a rapid method based on the former with hot NaOH solution was developed in 1987 by the Baotou Research Institute of Rare Earth. Using this method, the reaction time can be greatly shortened and NaOH consumption can be lowered, thus significantly decreasing production costs.

8.3.1.2. *Extraction of Rare Earths from Ion Adsorption Type Rare Earth Minerals*
The ion adsorption type of rare earth mineral was exploited only on a small scale from 1969-1979 and was not marketed, because the value of the product and the exploitation techniques were not sufficiently understood. Only in 1980, when a small quantity of the product was exported to Japan, did it begin to attract attention in rare earth circles at home and abroad. From then on it began to enter the international market. After 1985, rapid development of the exploitation and processing of this rare earth mineral occurred. In 1986 and 1988 3000 tons and over 8000 tons, respectively, were produced. In 1989, however, production fell. This type of mineral can be processed by a chemical method to obtain mixed rare earth oxides containing greater than 92% REO, or it can be further processed to obtain high purity individual rare earth oxides. The process is reliable and practical.

8.3.2. SEPARATION AND PURIFICATION OF INDIVIDUAL RARE EARTHS
Currently more than one hundred different rare earth products, mainly obtained by solvent extraction, ion exchange and oxidation-reduction methods are commercially available in China. Their purity ranges from 99% to 99.999%.

With improvements in rare earth extraction, the products made in China continue to increase in purity (Table 3). In 1980 the output of high purity rare earths was only 20 tons REO, while in 1988 it reached 1160 tons with an annual growth rate of 65%.

TABLE 3 **Production of High Purity Rare Earths in China 1986-1990 (t, REO)**

YEAR	OUTPUT OF HIGH-PURITY RARE EARTHS	TOTAL OUTPUT OF RARE EARTHS	RATIO (%)
1986	460	11860	3.9
1987	770	15100	5.1
1988	1160	18660	6.2
1989	1683	19670	8.6
1990	2109	14964	14.1

Source: China Rare Earth Factual Database

8.3.2.1. *Solvent Extraction Process*
Solvent extraction is the most important method in the world for extracting individual rare earths. Early in the 1970s Professor Xu Guangxian of Beijing University made a thorough and systematic study of the calculation theory for concatenated solvent extraction and optimal design, and proposed a series of practical mathematical models. As a result, the solvent extraction method has made considerable progress and a series of advanced processes have been developed. Some of these processes can be used to separate individual rare earths, and some are used for continuous separation of more than ten rare earth elements in the same medium, by using the same extractant.

Optimisation for product purity, recovery, stages of extraction and different parameters can be attained through calculation.

Extractants for rare earths are mainly of three groups: phosphines, amines and carboxcyclic acids. The largest quantities presently used in industry are phosphines. In the early 1950s Professor Yuan Chengye of the Shanghai Research Institute of Organic Chemistry, Academia Sinica and his assistants began devoting themselves to the study and synthesis of practical and effective extractants for rare earths, with a great many achievements. The equipment used for the extraction is mainly a mixer-settler. Extraction towers and centrifugal extractors are also used for some purposes. The mode of extraction is generally by the countercurrent and fractionation methods. In the late 1980s a more advanced technique - a three-outlet or a multi-outlet technique - was developed and has been used in production.

China leads the world in research and application of rare earth extraction technology and its history of development is almost as long as that of major rare earth producing countries.

8.3.2.2. *Other Separation Methods*
The ion exchange technique developed in the 1950s is an effective method and was once an important means for preparing individual rare earths with high purity in China. But on account of its long operation period and high cost, this technique has been gradually replaced by the solvent extraction method since the 1970s. However, it is still available for certain rare earth elements which are difficult to purify by extraction, or the purity of which has very strict requirements.

For certain rare earth elements with variable valency, like cerium and europium, production is carried out by oxidation-reduction.

8.3.3. PREPARATION OF METALS

China began preparing rare earth metals in the early 1950s for making lighter flints. With the development of the materials industry, fused salt electrolysis and metallo-thermic reduction were gradually developed in the 1960s and 1970s for preparing La, Ce and Sm etc. In the 1980s, as a result of the invention of Nd-Fe-B permanent magnets and the exploitation of yttrium for application in novel alloys, neodymium, yttrium, terbium and dysprosium were successively produced on an industrial scale. Now a variety of individual rare earth metals with a purity of 99% can be produced in the form of ingots, rods, powders, wires and strips. The preparation of high purity metals by processes such as zone smelting and solid state electromigration, etc. is still carried out on a laboratory scale. Should great quantities of these metals be demanded, these processes will be put into industrial operation.

8.3.3.1. *Fused Salt Electrolysis*
Fused salt electrolysis is the major process for manufacturing mischmetal and La, Ce, Pr, Nd and Y. Depending upon the composition of the electrolyte, the method can be divided into two large groups: the chloride system composed of $RECl_3$, KCl, NaCl, $CaCl_2$ etc. and the fluoride system composed of REF_3, BaF_2, LiF etc. The former is mostly used for producing mischmetal using $RECl_3$ as the raw material, whereas the latter is the main process for producing Nd and Y using RE_2O_3 as the starting material. The process is characterized by its continuous operation, high metal recovery and high current efficiency. It is, therefore, the major method for producing rare earth metals and alloys such as Nd-Fe and Y-Fe.

8.3.3.2. *Metallo-thermic Reduction*
Based on different reductants used, the metallo-thermic reduction method can be divided into two groups: La-(Ce-or mischmetal-) thermo-reduction and Ca-thermo-reduction. The former is mainly used to prepare metallic Sm, Eu and Yt. In comparison with La and Ce these metals have a lower boiling point and a higher vapour pressure, so during the course of the reduction of rare earth oxides by La thermo-reduction, Sm, Eu and Yt are obtained as vapours which are then condensed and collected to provide pure metals. The latter is suitable for preparing Gd, Tb, Dy and Y, etc. The metals obtained contain a small amount of Ca and other impurities which need further purifying.

To summarise, the processes for rare earth ore-dressing, extraction and separation are comparable to those used in developed countries,even though China started its rare earth industry later than these countries. However, compared with developed countries, China still lags behind in on-line analysis, automated control, steady production of high purity rare earths and analytical methods. For this reason, researchers have made great efforts to improve their technology so as to catch up with the rest of the world.

8.4. Application

In virtue of China's extremely abundant rare earth resources, the government and rare earth industrial circles have attached great importance to creating a large market for rare earth applications at home and abroad. The extension of their application at home, for example, has formed some special fields of application with national features. Over the past ten years the annual rare earth consumption at home has increased, on average, by 18.9%. In 1990 consumption totalled 7256 tons of REO, ranking first in the world. Tables 4 and 5 show the consumption of rare earth products in China and a comparison of their usage in China with that of the world's other major consumers of rare earths.

From Table 5 it can be seen that the consumption volume of rare earths in metallurgy, catalysis and glass and ceramics accounts for more than 85% of the total amount used both in China and abroad. As light rare earths are mainly used in these fields, they play an important role in balancing the consumption of light and heavy rare earths. The uses to which rare earths are put in China are not always the same as those in the Western world. For example, in China their application in glass and ceramics has developed only slowly. Although their consumption in new materials increases rapidly, the proportion remains very small. However, their application in agriculture, textiles and other light industries continues to rise quickly and presently accounts for over 10% of consumption.

The current RE-treated steel output has reached 250,000 tons per annum (Table 8), the most important of which are 16MnRE, 09CuPTiRE, 09MnRE, ZG20RE and GCr15RE, etc. mainly used in automobiles, bridges, pipelines, gears, bearings and various castings.

8.4.1. APPLICATIONS IN METALLURGY

Metallurgy was the earliest area in which rare earths were used in China. In the early 1950s RE-treated steel and nodular cast iron were studied and produced (RE-Mg master alloy was used as a nodulizer). The consumption of rare earths in this field has accounted for more than one-half of the total consumption. In recent years the consumption volume has continued to increase although the proportion has dropped somewhat (see Table 6).

The foundry industry consumes the largest amounts of rare earths. In 1950 China started developing nodular cast iron. The development of rare earth vermicular cast iron was initiated on the basis of a paper entitled "RE-containing high strength grey cast iron" published in 1966 by the Research Institute of Mechanical Design (Shandong Province). Now many spare parts for metallurgical machinery, automobiles, diesel locomotives and engines, etc., such as wearing plates, grinding balls, sludge pumps, carriers, ingot moulds, rolls, drawing heads, guides, pipes, cylinder casings, gears and crank shafts, etc., are made up of RE-treated nocular-, vermicular- and grey cast iron. The rare earth alloy is added mainly by pressing it into the molten iron. In recent years more advanced methods such as injection and converting ladle, together with their corresponding equipment, have replaced the old ones. The production of RE-treated iron castings now exceeds more than one million tons.

Since the beginning of the application of rare earths to steel making in the late 1950s, about 100 different steels have been tested. From the 1970s to the early 1980s research focussed on relieving or reducing teeming nozzle blockage for RE-treated steel and low multiple inclusion defects in steel bloom. Since the late 1970s a series of suitable methods for adding rare earths have been tried. These methods are listed in Table 7. The use of these methods has increased the recovery of rare earths in the steel and stabilized the steel quality which, in turn, help to promote constant expansion of the practical limits for RE-treated steel.

TABLE 4 Consumption of Rare Earths in China 1978-1990 (t, reo)

YEAR	TOTAL	METALLURGY	PETRO-CHEMICALS	GLASS & CERAMICS	HI-TECH	OTHERS*
1978	1000	900	100	--	--	--
1979	1050	950	100	--	--	--
1980	1724	1235	406	83	--	--
1981	1628	941	440	80	5	158
1982	2019	1243	555	100	20	101
1983	2505	1765	580	130	30	--
1984	3000	2108	580	160	32	120
1985	3500	2250	700	350	40	160
1986	4222	3022	700	200	40	260
1987	4888	3240	950	250	50	398
1988	6000	3410	1600	300	70	620
1989	6770	3500	2030	360	80	800
1990	7256	3600	2200	410	95	951

* - including agriculture, textile and other light industries

Source: China Rare Earth Factual Database

TABLE 5 Consumption Of Rare Earths In Metallurgy 1986-1990 (t, REO)

PRODUCT	1986	1987	1988	1989	1990
Iron	2700	2870	2780	2730	2800
	(64)	(58.7)	(46.3)	(40.3)	(38.6)
Steel	212	220	300	370	400
	(5)	(4.5)	(5)	(5.5)	(5.5)
Non-ferrous	110	150	330	400	400
metal	(2.6)	(3.1)	(5.5)	(5.9)	(5.5)
TOTAL	3022	3240	3410	3500	3600
	(71.6)	(66.3)	(56.8)	(51.7)	(49.6)

Figures in brackets indicate the percentage of rare earths constituting the total consumption

Source: China Rare Earth Factual Database

TABLE 6 Methods Of Adding Rare Earths To Steel

YEAR	METHOD
1976-1980	Suspending rare earth metal rod in a steel mould
1979-1981	Pressing RESiFe alloy in a ladle
1981-1984	Rare earth wire feeding in a crystalliser for continuous slab casting
1982-1984	Rare earth wire feeding in a git
1982-1987	Injecting RESiFe powder into solid synthetic shielding slat in a ladle
1983-1987	Adding RESiFe powder to pouring steel stream
1986-1988	Suspending rare earth metal rod in massive mould
1987-1990	Injecting RESiFe powder into a git
1988-1990	RESiFe powder cored wire feeding

Source: China Rare Earth Factual Database

TABLE 7 Output Of Re-Treated Steel (t)

YEAR	OUTPUT
1980	15685
1981	29854
1982	69787
1983	88359
1984	114012
1985	111553
1986	140800
1987	165561
1988	245049
1989	256980

Source: China Rare Earth Factual Database

The use of rare earths in Al-Si alloy for making pistons in the 1960s initiated the application of rare earths in the non-ferrous industry. In the 1970s rare earths were used in high tension Al-Mg-Si alloys for cables and in Fe-Cr-Al alloys for electric heating wire. Between 1980 and 1985, RE-Al alloys were first obtained by low temperature electrolysis, and were later produced directly by adding rare earth compounds to an Al electrowinning cell. The production of RE-containing Al cables has developed rapidly because of its excellent electrical conductivity and workability. The alloy has been tested stepwise from 10kV to 220kV and proved to be in accord with the IECS's standard.

Rare earths are also used in conjunction with other non-ferrous metals like Mg, Zn, Cu and Ti, etc. with satisfactory results. The substitution of RE-containing zinc alloy for zinc in a coating provides excellent corrosion resistance. In recent years the application of rare earths in silver contact materials has also occurred to good effect. The increasing rate of application of rare earths in non-ferrous metals (mainly Al) since 1986 is evident (Table 6).

8.4.2. APPLICATIONS IN THE PETRO-CHEMICAL INDUSTRY

RE-containing zeolitic cracking catalysts consume the largest amount of rare earths in the petroleum industry. The use of this catalyst began in 1960s but did not become widespread until 1980 by which time RE-containing catalysts made up about 1/3rd of the total catalysts used in all the oil refineries. Now almost all the catalysts used in this field have been substituted by RE-containing catalysts (Table 9). Although superstable zeolitic catalysts of type Y without rare earths or with only a small amount of rare earths are being tested and extended, it is believed that it will not be able to replace the former in the near future.

In the chemical industry, rare earths are used as a drying agent in paint, as auxilliaries in synthetic ammonia, synthetic rubber and plastics, and as a catalyst for exhaust gas purification. These products were studied in the 1970s and have been put into operation successfully in the 1980s. Now a variety of RE-containing catalysts for exhaust gas purification have been produced on a large scale. The problem of purifying black smoke from diesel engines has also been studied.

TABLE 8 Proportion of RE-Containing Zeolitic Catalyst in Total Catalyst (%)

YEAR	1980	1982	1984	1986	1988
RE-containing zeolitic catalyst	34	68	81	95	95
High-activity RE-containing zeolitic catalyst	17	28	56	76	86

Source: China Rare Earth Factual Database

8.4.3. APPLICATIONS IN GLASS AND CERAMICS

In the traditional glass and ceramics industry, rare earths are mainly used as polishing, decolouring, and colouring agents for glass, and as colouring agents and glazes for ceramics. They are also used for making special glass, etc.

The following achievements mark the beginning of the application of rare earths in glass making in China: (1) the development of high cerium polishing powder in 1974 made by Shanghai Yuelong Non-ferrous Metals Company (formerly Shanghai Yuelong Chemical Plant); (2) the use of CeO_2 and mixed rare earth oxides as a decolorant and defecating agent in the early 1960s by Wuham Glass Instrument Factory; and (3) scientific research on the "Effect of Ceric Ion on Colourization of Titanium Glass" carried out in the mid-1960s by the Shanghai Institute of Silicate, Academia Sinica. Research on rare earth optical glass was made by the Changchun Institute of Optical Device and the Beijing Research Institute of Glass and Ceramics in 1964-1965. Three kinds of RE-containing colourless optical glass with 28 trademarks are now produced in China. They are the lanthanum crown glass (LaK_1 - LaK_{12}), lanthanum flint glass (LaF_1 - LaF_{10}) and heavy lanthanum flint glass ($ZLaF_1$ - $ZLaF_4$). Ten million eye lenses (generally containing cerium) are produced annually by No. 603 Factory in Beijing. The properties of the lens have attracted international attention. This product is partly marketed at home and is partly sold abroad. In the 1980s the following products were developed and production has been conducted to varying scales: neodymium-doped silicate and phosphate laser glass, cerium- and terbium-magnetic glass, lanthanum-, gadolinium- and yttrium-fluoride glass fibre, ytterbium-, erbium-, cerium-, and some other rare earth luminous glass, gadolinium-, europium-, dysprosium-, etc. radiation-proof glass and cerium-containing radiation-resistant glass.

China is world-known as a porcelain producing country. In the 1970s, different kinds of pigments known as praseodymium-yellow, praseodymium-green and neodymium-violet, etc. were applied to ceramics with satisfactory effects; RE-coloured sanitary ware is exported to over 50 countries and regions; high temperature neodymium glaze with double-colour effect appears violet under sunlight and blue under fluorescent light; low temperature cerium glaze with high emulsibility may distinctly increase the whiteness and lustre of glazed tiles, and rare earths can also be used as a suspending agent for enamel glaze.

Table 5 also shows recent rare earth consumption for conventional uses in the glass and ceramics industry. Though the application area has been constantly expanded, no significant increase in the use of rare earths in polishing powders, which are large rare earth consumers, has been made as it is less competitive. The annual rate of increase in the use of rare earths in this field is close to that of the total consumption, making up a constant 5-6% of the total.

8.4.4. APPLICATIONS IN HIGH-TECHNOLOGY FIELDS

The application of rare earths in high-technology includes their use in magnets, fluorescent materials, lasers, and fine ceramics, etc, in which rare earth elements are either the principal constituent or micro-constituents in the form of additives.

Since the 1960s, Chinese investigators have begun researching into rare earth permanent magnets, phosphorescent materials, hydrogen storing materials and sensing materials, etc. Once a new material came into being in foreign countries, the same matrial would almost simultaneously be developed in China. The development of the well-known rare earth superconductor is a good example. Presently the research level and production of rare earth materials in China are comparable with those in developed countries.

Rare earth permanent magnetic materials are the largest rare earth consumer in the field of high-technology in China. Scientific research institutions engaged in research on permanent magnet materials mainly include the Central Research Institute of Iron and Steel, Beijing; the Baotou Research Institute of Rare Earth; General Research Institute of Non-ferrous Metals, Beijing; and the Shanghai Research Institute of Iron and Steel. Many of their achievements have been close to or have attained the world's most advanced standards. The prototype of NdFeB permanent magnet made by Baotou Research Institute of Rare Earth in 1990 has attained a magnetic energy product of $415kJ/m^3$, which is the highest that has ever been reported. Now studies on PrNdFeB permanent magnets with high operating temperatures and low temperature coefficients and other new types of magnets are under way. In China there are tens of rare earth permanent magnet producers producing these magnets for domestic demand and for export. The substitution of rare earth permanent magnets for traditional ones used in motors, gyroscopes, anti-wax devices and dehydrators in oil mining has prompted producers to develop a number of different products, with satisfactory results.

Rare earth luminescent materials which have been developed in China mainly include photoluminescence-, cathodoluminescence- and X-radioluminescence materials. As a result of 20 years' laborious and tiresome study, rapid progress has been made in the research and production of luminescence material. The rare earth trichromatic phosphorescent energy saving lamp is an example, which is characterized by its high luminous efficacy, distortionless colour and energy saving. It is now being actively marketed. The trichromatic phosphor is composed of red (Y_2O_3:Eu), green [(Ce,Tb)MgAl$_{11}$O$_9$] and blue [(Ba,Eu)Mg$_2$Al$_{16}$O$_{27}$] phosphor and was first developed domestically by Fudan University in 1980. More than ten producers are now manufacturing this material. The annual production capacity has reached 50 tons, but in 1990 the actual output was only 23 tons. Shanghai Yuelong Non-ferrous Metal Company, the major producer of this material, is now extending its production line for further increasing production. The production capacity of trichromatic lamps has reached 30 million. In 1990 the actual output was 18 million, part of which was exported. The cathodoluminescence materials consisting of rare earths as a matrix mainly include three systems: Ln_2O_2S, $Y_3M_5O_{12}$ and LnOX. Engaged in developing these materials are the Changchun Institute of Applied Chemistry, Academia Sinica, Changchun Institute of Physics, Academia Sinica, Fudan Unviersity and Beijing University among others.

Fine ceramic materials including engineering and electronic ceramics are also under investigation in China. The successful development of a ceramic engine demonstrates that the development of engineering ceramics in China has kept pace with the rest of the world.

Other functional materials such as LaNi$_5$ or MMNi$_5$ alloy used in Ni-H batteries, single crystal thin film of (BiTm)$_3$ (FeGa)$_5$O$_{12}$ and (BiPrGdYb)$_3$ (FeAl)$_5$O$_{12}$ used in magneto-optical modulators, Gd$_3$Ga$_5$O$_{15}$ monocrystal used as substrate materials and rare earth superfines, etc. are being developed and encouraging progress has been made.

8.4.5. OTHER APPLICATIONS

The application of rare earths in agriculture, textiles and other light industries is a new domain for rare earth application opened up in China with distinctive features.

Study of the utilization of rare earths in agriculture began in China in 1972. A great deal of experience has shown that the appropriate application of rare earths to crops can increase production and improve quaity. The RE-containing fertilizer used for this purpose is one that contains as an additive a small amount of a soluble rare earth compound. If applied by the technique of seed dressing or sprinkling over the leaves, an increase in production by 5-15% can be generally expected. Toxicological investigation has proved that this fertilizer is of low toxicity and is not a teratogenic or carcinogenic substance. Experiments also show that only a very small amount of it is absorbed by plants. Compared with non-RE-treated wheat, the rare earth content and radioactivity of wheat to which RE-fertilizer has been continually applied for nine years have not shown any obvious change. Instead, rare earth concentrations remain within the background level present naturally. The land used for applying RE-containing fertilizers has gradualy increased in the past 10 years, reacing 30 million mu (2 million hectares) in 1990. Rare earth fertilizers show similar growth providing effects on grass, forest trees, flowers and plants.

The technique of using rare earths as a dyeing auxilliary was first developed by No. 1 Woollen Mill of Inner Mongoliain 1977. This technique developed after 10 years' constant research, can now be used to treat natural fabrics such as pure wool, hemp, silk and cotton; synthetic fibers such as nitrilon, chinlon and nylon, as well as fur and leather. It produces the following results: it gives a bright and lustrous colour; it increases the dyeing rate; it economises on dyestuff and labour and it reduces environmental pollution. Leather tanning in the presence of rare earths may also improve product quality and reduce environmental pollution. All these techniques have been extended over the whole country with remarkable economical benefit.

8.5. Exportation

China exported only a small amount of its rare earth products in 1973. After entering the eighties, however, China has become an important rare earth exporting country (see Table 10). Among important users are Japan, the United States and western European countries. The exports have also been changed from the early crude minerals to processed products. This indicates that both the ability and level of rare earth processing in China have greatly improved.

TABLE 9 Export Volume and Value

YEAR	1985	1986	1987	1988	1989	1990
Volume (t,REO)	4500	5187	6500	8320	9154	6139
Value ($x10^{-6}$)	38	43.2	61.6	102.7	118.6	83.5

Source: China Rare Earth Factual Database

China's abrupt rise in its status as a major producer, consumer and supplier of rare earths and rare earth products is the most important event of the 1980s in terms of development of rare earths. The six year plan (1990-1995) currently mapped out by the government is focussing on domestic applications. By the end of 1995 an 8% increase for China's demand for rare earths can be expected. To be optimistic, China's rare earth industry will have a great future and will occupy an even more important position in the world's rare earth industry.

CHAPTER 9

RARE EARTH ELEMENTS IN THE GEOLOGICAL SCIENCES

EDWARD G. LIDIAK
Department of Geology and Planetary Science
University of Pittsburgh
Pittsburgh, Pa. U. S. A. 15260

WAYNE T. JOLLY
Department of Earth Sciences
Brock University
St. Catharines, Ontario, CANADA L2S 3A1

9.1. Introduction

Rare earth elements (REE) are an integral part of modern geochemical and petrological studies. The rare earths are widely used, either alone as a group or in combination with other trace elements, to study igneous rock systems and, to a lesser extent, sedimentary and metamorphic rocks. Most of the major advances have been in the field of igneous petrogenesis and geochemistry, where they have been particularly important in evaluating the composition and history of magmatic source regions, the conditions of melting, the extent of melting, the modification of melt composition by assimilation or metasomatism, and the magmatic differentiation processes of the resulting magmatic system.

From an historical standpoint, the use of the rare earths in geochemical analysis and modeling is a fairly recent development. Prior to the 1960s, REE were rarely used in geochemical studies. Since that time there has been a literal explosion in the systematic use of REE in geochemistry. Clearly, the main reasons for this change in approach to geochemical problem solving has been the improvement in analytical methodology and precision and the development of quantitative approaches, thereby allowing enhanced interaction among fact, models, and theory.

A second important impetus in the use of REE in geochemical analyses was the advent of plate tectonic theory. Although the development of REE systematics and the concepts of plate tectonics are casually related at best, their quantitative development in the last three decades has been a happy and successful marriage. The rare earth and

149

C. H. Evans (ed.), Episodes from the History of the Rare Earth Elements, 149–187.
© 1996 *Kluwer Academic Publishers. Printed in the Netherlands.*

other trace elements, for example, are particularly useful in recognizing and characterizing modern tectonic environments and in quantifying geologic hypotheses.

Finally, the utility of REE in petrogenetic studies derives from the fact that, although the rare earths are a genetically related group of elements, their partition coefficients (ratio of the concentration of an element in a mineral to its concentration in the coexisting liquid) vary systematically with atomic number and differ from mineral to mineral (McKay, 1989). This difference results in strikingly different concentrations and patterns in rocks (comprised of one or more minerals), and it is commonly possible to identify which minerals have been involved in the formation of a particular rock. Furthermore, the use of Sm-Nd isotopic systematics, and to a lesser extent Lu-Hf and La-Ce, has proven to be a powerful tool in elucidating petrogenetic processes and in the timing at which various geologic events occurred.

9.2. Scope

The paper is not intended to be a comprehensive literature review of the use of the rare earths in the geological sciences. Its purpose, instead, is to provide an historical perspective of the contributions that geochemists have made to the understanding of rare earth element behavior and to discuss some recent developments. The first part of this review deals with historical developments; the second part considers specific examples of how REE are currently used in the geological sciences.

Earlier, excellent review volumes and articles on the geochemistry of REE have been published by Haskin and Frey (1966), Haskin and others (1968), Allegre and Hart (1978), Allegre and Minster (1978), Haskin (1979), Haskin and Paster (1979), Hanson (1980), Henderson (1984), and Lipin and McKay (1989) (and including articles and references, therein).

9.3. Geochemical Characteristics

The rare earth elements form a coherent group of 15 elements from La (Z=57) to Lu (Z=71) within Group III of the Periodic System; fourteen of these occur naturally. The elements are geochemically very similar, exhibiting strong lithophile tendencies and, with the exception of Eu and Ce, being trivalent under most geologic conditions. The series forms a notable exception to the regularity of the periodic table in that, in general, the ionic radius increases with increasing atomic number if the ionic charge and the number of electrons in the outermost shell remains unchanged (Rankama and Sahama, 1950). Rather, in the lanthanide series, the radii of the trivalent ions in octahedral coordination decreases systematically with increasing atomic number (Fig. 1), an effect known as the lanthanide contraction. This behavior stems from the fact that the REE contain two electrons at the 6s energy level and one electron at the the 5d level. With increasing atomic number, additional electrons are added to the 4f energy level in the inner N shell rather than to the outer electron shells. Thus, La has no electron at the 4f level, Ce has one, Gd has seven, and Lu has 14.

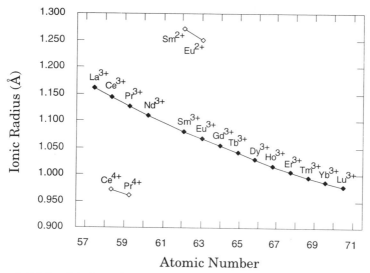

Figure 1: Relation of atomic number to ionic radius in the rare earths (lanthanide contraction). Additional valence states for several of the elements are shown for comparison. Except for Eu, REE are trivalent in most geologic environments. Ionic radii (8-fold coordination) are from Shannon and Prewitt (1970) and Shannon (1976).

A given REE is geochemically very similar to its nearest atomic neighbor, but is systematically different from those REE with larger or smaller atomic numbers. In general, REE with lower atomic number, the light rare earths (LREE), behave differently and are geochemically distinct from the intermediate or medium rare earths (MREE) and these differ from the heavy rare earths (HREE). REE concentrate mainly in minerals having appropriate cationic radii sites. The LREE favor relatively large radii (about 1.0 Å) similar to Ca^{2+} and Th^{4+}, whereas the HREE prefer smaller sites approaching that of Mn^{2+} (Hanson, 1980; Burt, 1989). Calcium-bearing minerals such as augite, hornblende, and garnet typically are enriched in the rare earths, whereas a mineral such as olivine is depleted (Fig. 2). Similarly, the LREE/HREE ratio is high in K-feldspar and low in hypersthene. It is also important to note that the REE content of an igneous mineral is controlled by the partitioning effects (coefficients) between a mineral and the liquid from which it crystallized and by the concentration of REE in the source. Partition coefficients of REE in some common minerals is shown in Figure 3. Olivine has a very small partition coefficient (<0.03) for all of the REE. The REEs thus tend to concentrate in the liquid rather than in the coexisting olivine during melting or crystallization; however, the overall concentrations are so low (Fig. 2) that olivine is incapable of producing significant REE fractionation. In contrast, garnet has high concentrations of REE, particularly the HREE. The partition coefficients for garnet span a large range, with the LREE being slightly incompatible and the HREE being compatible. During a magmatic event, the LREE would concentrate in the liquid and the HREE would concentrate in the solid.

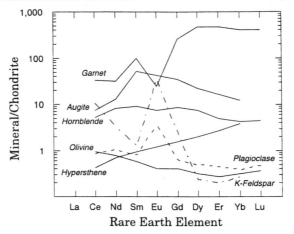

Figure 2: Chondrite-normalized REE diagram of common rock-forming minerals. Mineral data from Schnetzler and Philpotts (1970)

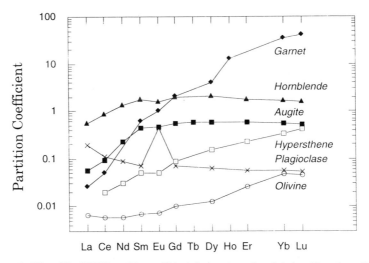

Figure 3: Mineral/liquid REE partition coefficients for common minerals in basaltic systems. Data from (Rollinson, 1993).

Graphical plots of REE in various rocks and minerals are generally normalized to chondritic meteorites or some other standard parameter. Normalization is done to eliminate the effects of the Oddo-Harkins rule (even numbered elements are more abundant than those with odd atomic number, Fig. 4) and to allow direct comparison to a known standard. Chondrites are used as a normalizing factor because they represent primitive solar material that may have been part of parental earth. In this paper, the chondrite concentrations of Sun and McDonough (1989) are used unless otherwise stated.

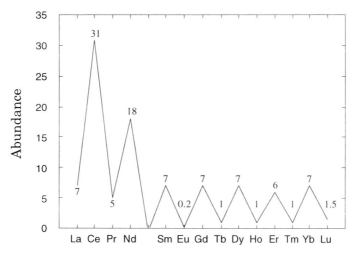

Figure 4: The relative abundances of the rare earths normalized to yttrium = 100 (from Goldschmidt and Thomassen, 1924)

9.4. Early Historical Developments

The term *rare earth* is somewhat of a misnomer. These elements were originally separated and identified as earths (heavy oxides) from relatively rare minerals, and they were thus characterized as rare earth elements. However, they are at least as abundant in the earth's crust as the more common minor elements such as Li, Cs, Pb, Th, and U (Goldschmidt, 1937; Taylor and McLennan, 1985). Further, they are not restricted to rare minerals, but typically occur as dispersed elements in a wide variety of minerals and rocks. They have been aptly characterized as being dispersed and not-so-rare (Haskin and Frey, 1966).

As described by Pyykkö and Orama in Chapter 1 of this volume, the rare earths were first discovered in 1794 by Johan Gadolin who separated and analyzed new "earths" from a mineral specimen collected by C. A. Arrhenius a decade earlier from a quarry in Ytterby, Sweden (Vickery, 1953, 1961). The new mineral was given the name *gadolinite* and the new earth was called *yttria*. Thus began a fascinating realm of chemistry (and geochemistry) that embraces a series of naturally-occurring elements of extraordinary similarity (Moeller, 1963).

The initial discovery was followed in 1803 by the independent identification of a similar and yet somewhat different heavy earth from another new mineral by Klaproth and by Berzelius and Hisinger (Moeller, 1963). This element was named *ceria* and the mineral *cerite*. A detailed account of this subject is provided by Trofast in Chapter 2 of this book. These initial discoveries were followed by the critical investigations in 1839-1841 by C. G. Mosander, a Swedish surgeon, chemist, and mineralogist, and former assistant to Berzelius. Mosander proved that both *yttria* and *ceria* are complex oxides rather than individual elements. {In modern terminology, ceria would include minerals rich in the LREE (La-Sm) and yttria minerals rich in the HREE (Eu-Yb)}. He also succeeded in isolating the first rare earth, *lanthana*. Mosander's work is described in greater depth by Tansjö in Chapter 3 of this volume. It was not until the second decade of the 20th century, when the association of K-series of X-ray spectra with atomic number was developed by Mosely, that the number of elements for the group was finally established (Haskin and Frey, 1966).

The close chemical similarity of REE compounds required tedious crystallization, precipitation, thermal decomposition, and extraction procedures, all of which had to be fractional in character. Until comparatively recently, four main methods were available: group methods for separating LREE from HREE; precipitation methods for isolating Ce and Eu; amalgam extraction for isolating Sm, Eu, and Yb; and fractional crystallization for separating the remaining elements (Topp, 1965). These procedures were replaced by ion-exchange and solvent extraction techniques.

The early prominent geochemists (Clarke, 1924; Clarke and Washington, 1924; Fersmann, 1932; Goldschmidt, 1937) were concerned mainly with three main basic problems of geochemistry: (1) obtaining analytical data on a variety of rocks and estimating the concentration of elements in the main subdivisions of the earth and in meteorites; (2) determining the general laws and principles that underlie the relations between the elements of the periodic table and their concentrations in nature; and (3) determining concentrations of elements in minerals for the purpose of evaluating ore deposits (Allegre and Minster, 1978). In these and other early works on elemental abundances, Ce, La, Yb and the related Y are the only rare earths that are commonly mentioned as being present in minor amounts in a variety of rocks; however, abundance figures were generally not given.

A remarkable aspect of this early geochemical work was the establishment of many of the laws governing the distribution of elements in crystalline substances. After the end of World War I, Goldschmidt and his associates began an extensive and systematic study of the abundance, distribution, and behavior of the REE. At that time, reliable data on the REE were almost totally lacking due to the difficulty in separating individual elements by classical chemical methods. However, a significant breakthrough occurred in the use of X-ray spectra to identify the individual REE; a single spectrum revealed the characteristic lines of each individual element. The X-ray spectra, when considered as part of a coherent group, display a relatively simple pattern of lines in regular progression from one element to the next and, moreover, the intensity of the lines is approximately proportional to the amount present (Mason, 1992). In addition to demonstrating the spectral characteristics of the REE, Goldschmidt and Thomassen (1924) estimated the relative abundances of these elements. Figure 4, adapted from Goldschmidt and Thomassen (1924), shows the relative abundances of the REE and confirms that Ce, followed by Nd, is the most abundant of the rare earths. Figure 4 also shows the rather remarkable difference in abundance between even- and odd-numbered elements. Goldschmidt and Thomassen (1924) were able to formulate precisely what is now know as the Oddo-Harkins rule: "Elements of odd atomic number are less abundant that their immediate neighbors of even atomic number" (Mason, 1992).

Goldschmidt and his assistants (Goldschmidt and others, 1925) also studied, using X-ray diffraction techniques, the relations of ionic size to crystal structure of the REE oxides. This work led to the remarkable discovery that the unit cell dimensions of the REE (lanthanide) oxides decrease regularly with increasing atomic number, a phenomenon he (Goldschmidt and others, 1925) referred to as the *lanthanide contraction*. He also published the first table of atomic radii of the elements (Goldschmidt and others, 1926). One of Goldschmidt's associates, Minami (1935), was the first person to publish measured REE concentrations in common rocks (shales). Goldschmidt went on to make numerous contributions to the distribution of elements in minerals and rocks, to geochemical abundance studies, and to geochemical principles (Mason, 1992). Beginning in 1929, he converted to the optical spectrograph in order to investigate the abundance and distribution of elements at very low concentrations, developing techniques including the use of internal standards, that resulted in dramatic improvements in precision and accuracy (Strock, 1936).

Following World War II, improved methods of spectrographic analyses (Mitchell, 1938, 1948; Smith and Wiggins, 1949; Ahrens, 1950) permitted the detection of a large number of elements with only a minimum of chemical treatment. Most analytical results were reported to an accuracy of about ±10-30 %. The new emphasis, led by researchers in Great Britain (Wager and Mitchell, 1951, 1953; Nockolds and Allen, 1953, 1954, 1956), focused on geologically well-documented suites of igneous rocks. REE did not play a prominent role in these initial studies as only La and Y were reported in the analyses. This research was, however, significant as it demonstrated to a large geological audience the potential of trace element studies in genetically related groups of rocks.

During this period, a cooperative international investigation of analytical precision and accuracy was initiated using two U. S. Geological Survey standards: diabase, W-1, and granite, G-1 (Fairbairn and others, 1951; Rollin and others, 1960). These inter-laboratory studies reflected the state-of-the-art in geochemistry at the time. The generally poor comparative results obtained for many elements by independent laboratories were distressing but nevertheless illuminating to most scientists. As an example, three separate analyses of La, the only REE reported in the first study, yielded the following values for G-1 (in weight percent La_2O_3): 0.050, 0.022, and 0.015 (Ahrens, 1951). The high and low values differ by a factor of 3! During this time, statistics were also utilized to help evaluate accuracy and precision (Ahrens, 1954a,1954b, 1954c, 1957), but with limited success (Chayes, 1954; Miller and Goldberg, 1955). A decade later, somewhat improved results were obtained for La and Yb, but a considerable range in concentration was reported for Ce (Berman, 1957; Ahrens and Fleischer, 1960). These elements were analyzed mainly by spectrochemical techniques. Other minor and trace elements were analyzed by a variety of methods, including chemical, spectrochemical, X-ray fluorescence, neutron-activation, and isotope dilution. It is perhaps not surprising in hindsight that the most accurate results were obtained by the latter two procedures. In many respects these results effectively mark the end of an era in trace element and REE geochemistry where the old methods of optical spectroscopy gave way to more precise methodology.

9.5. Modern Era

The two analytical techniques that came to the fore in the 1960s (Allegre and Hart, 1978) were isotope dilution mass spectrometry, the standard method of isotope geochemistry, and neutron activation analysis. As applied to trace element and REE geochemistry (Gast, 1960, 1968; Masuda, 1962; Schnetzler and Philpotts, 1970; Arth and Hanson, 1975; Kay and Gast, 1975; Sun and Hanson, 1975), isotope dilution remains a highly accurate and precise technique. The second technique, neutron activation analysis (INAA), was initially applied to meteorite studies (Goldberg and others, 1951; Smales and others, 1957; Schmitt and others, 1963; Gordon and others, 1968) and then extended to terrestrial rocks--for example, Haskin and Gehl (1963). INAA analysis continues to be used as a precise analytical method. Recent advances and summaries of modern techniques and methodologies are described by Thirlwall (1982) and Henderson and Pankhurst (1984). A third procedure, inductively coupled plasma (ICP) mass spectrometry (Longerich and others, 1990), is increasingly being used today as it permits the accurate analysis of all 14 REE.

According to Allegre and Minster (1978), modern trace element geochemistry was initiated by the simultaneous contributions of Masuda (1962) in Japan and Winchester and co-workers (Coryell and others, 1963; Towell and others, 1965; Schilling and Winchester, 1966) in the United States. It is noteworthy that Masuda used isotope dilution and Winchester used INAA as their analytical methods. Of particular significance is the fact that both independently proposed the use of chondrite-normalized abundance patterns in evaluating REE concentrations and behavior. The utility of this approach is that it readily allows comparison with an established standard (in this case the average chondrite meteorite), shows the concentration relative to the reference, reveals the regularity of most REE abundance patterns, and permits clear comparison of co-genetic samples. These studies marked the beginning of modern quantitative trace element geochemistry.

Perhaps the single most important individual influence exerted on modern trace element geochemistry was by Paul Gast. Gast's (1968) contribution was threefold (Allegre and Minster, 1978): (1) he introduced in partial melting models the mineral/liquid partition coefficient concept developed previously for fractional crystallization (Neuman and others, 1954); (2) he integrated isotope and trace element studies; and (3) he used isotope dilution in trace element analysis. Gast believed that trace elements (including the REE) were important in understanding igneous petrogenesis. Most discussions of the problem of parental magmas and the genetic relations between magma types at that time were in terms of major element bulk compositions and phase equilibria relations (Kuno, 1959; Yoder and Tilley, 1962; O'Hara, 1965; Green and Ringwood, 1967). Gast recognized that trace elements carry important information bearing on the genetic relation among magmas and co-existing solids. He noted that the concentration of a trace element in a volcanic liquid depended on three factors: (1) the concentration of the element in the source region, (2) the extent of chemical fractionation that occurs during melting, and (3) the amount of chemical fractionation that takes place during ascent and crystallization of the liquid. With this in mind, he formulated several crystallization and melting models that quantified the changes in trace element concentrations during partial melting. These equations were later reformulated by Shaw (1970). A summary of these quantitative models is given by Cox and others (1979) and derivations of the basic equations are provided by Wood and Fraser (1978).

9.6. Recent Developments

During the 1960s and continuing today, quantitative trace element geochemistry and the plate tectonic paradigm experienced parallel and explosive development. These developments were, at least in part, complimentary. This relation can perhaps best be demonstrated with reference to oceanic basalts. Prior to plate tectonics concepts, it was generally assumed that ocean island basalt formed the bulk of the oceanic crust and with ocean ridge basalt being greatly subordinate (Green and Poldervaart, 1955). Basalts were distinguished on the basis of silica saturation (Yoder and Tilley, 1962) into tholeiite basalt (silica saturated) and alkali basalt (silica undersaturated), but were not characterized according to tectonic association.

With the advent of plate tectonic theory (Hess, 1962), and specifically the recognition that magnetic polarity reversals along oceanic ridges are related to basaltic volcanism and sea floor spreading (Vine and Matthews, 1963), the volumetric abundance of basalt from oceanic ridges became apparent (Kay and Hubbard, 1978). Chemical evidence available at that time (Frey and Haskin, 1964; Engel and others, 1965) indicated that ocean ridge basalts are distinct in containing low amounts of Ba, K, P, Pb, Rb, Sr, Ti, Zr, U, Th and the LREE. Although not specifically emphasized, these workers noted that there are significant differences between ridge and island basalts.

The recognition of significant geochemical differences between shield-building tholeiites on Hawaii and ocean ridge tholeiites (Hubbard, 1967; Schilling and Winchester, 1969) led to the realization that they are erupted in different tectonic settings. A new nomenclature consistent with plate tectonic theory thus evolved in which mid-ocean ridge basalt (MORB) characterizes divergent, constructive plate margins and ocean island basalt (OIB) erupted through thickened oceanic lithosphere characterizes oceanic intraplate environments.

9.6.1. MID-OCEAN RIDGE BASALT

Volcanism along a constructive plate margin such as the mid-ocean ridge system should be among the simplest type of basaltic magmatism as the vast outpouring of magma occurs in a related tectonic setting that is removed from the possible influence of continental crust. Early studies of MORB suggested that, perhaps, there was general compositional uniformity. There are, however, distinct variations along the length of the ridge system where two main end-members or extreme types of tholeiite basalt are erupted, a normal type (NMORB) and a plume or enriched type (EMORB). Most NMORB typically occurs along normal segments of the ridge, whereas EMORB is mainly restricted to shallow ridge platforms such as the Azores and Iceland (Schilling and others, 1983). NMORB (Fig. 5) is characterized by strong depletion in LREE and essentially unfractionated HREE patterns (Frey and others, 1974; Sun and others, 1979; Wood and others, 1979; Schilling and others, 1983). In contrast, E-type MORB (Fig. 6) is more enriched in LREE and has similar levels of HREE (Humphris and Thompson, 1983; Le Roex and others, 1983; Schilling and others, 1983). The more primitive NMORB and EMORB have LREE concentrations of about ≤10x chondrites and ≤30x chondrites, respectively. Higher concentrations reflect fractionation effects. The distinction between NMORB and EMORB stems from differences in magma source areas and crystallization history (Le Roex, 1987). NMORB derives from a depleted upper mantle source in which partial melting was fairly extensive. In contrast, the source of EMORB is a more enriched or plume-enhanced upper mantle.

9.6.2. OCEAN ISLAND BASALT

A second type of oceanic basalt characterizes oceanic islands. Oceanic islands are a prime example of a long-lived thermal anomaly (hot spot) in an intraplate tectonic setting. The Hawaiian Islands, part of the Hawaiian-Emperor seamount chain which have been in existence during the past 70 Ma, are the most prominent example. Most OIB are distinct from EMORB in REE concentrations (Fig. 7). The data shown in Figure 7 compare only tholeiitic basalts in both tectonic settings so that the differences primarily reflect variations in source compositions (Jones and others, 1993).

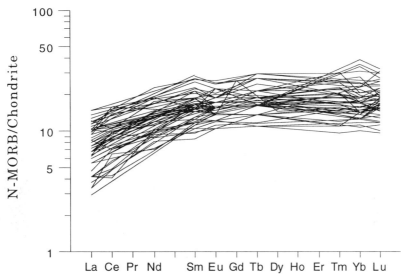

Figure 5: Chondrite-normalized REE distribution in N-type ocean ridge basalt (NMORB). Data sources: Frey and others (1974), Sun and others (1979), Wood and others (1979), and Schilling and others (1983)

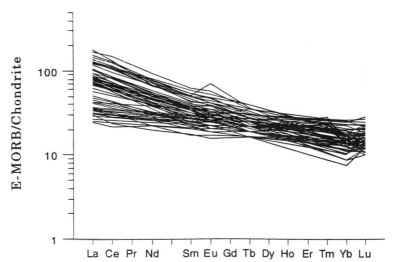

Figure 6: Chondrite-normalized REE distribution in E-type ridge basalt (EMORB). Data sources: Humphris and Thompson (1983), LeRoex and others (1983), and Schilling and others (1983)

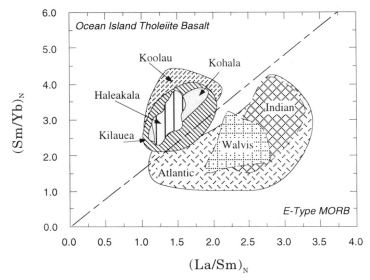

Figure 7: Discrimination diagram of La/Sm vs. Sm/Yb (both normalized to chondrites) showing distinction between ocean island tholeiite basalt and E-type MORB. Straight dashed line empirically separates the two main fields (Jones and others, 1993). Data sources for EMORB: Atlantic Ocean (Schilling and others, 1983); Indian Ocean (Le Roex and others, 1983); Walvis Ridge (Humphris and Thompson, 1983). Data sources for ocean island tholeiite: Haleakala volcano (Chen and Frey, 1985); Kilauea volcano (Tilling and others, 1987); Kohala volcano (Lanphere and Frey, 1987); Koolau volcano (Roden and others, 1984; Frey and others, 1994).

Each of the Hawaiian shield volcanoes evolved in stages (Macdonald and Katsura, 1964; Clague and Dalrymple, 1987). The main stage, forming 95-99% of the volcano, is the development of a tholeiitic shield. The voluminous tholeiites are followed by a relatively thin veneer of post-shield tholeiitic, transitional, and alkalic basalts (Frey and others, 1990). The final post-erosion stage is characterized by minor outpourings of highly alkalic and nephelinitic lavas. The latter two stages are not developed on every volcano. The shield stage may be followed by a post-shield stage, a post-erosion stage, or by both.

One of the many intriguing aspects of Hawaiian volcanism is the temporal evolution from tholeiitic basalt through alkalic basalt to highly undersaturated alkalic basalt.

These changes involve not only the major elements, but also trace elements and REE. The REE patterns were first recognized by Schilling and Winchester (1966). Systematic studies (Feigenson and others, 1983; Chen and Frey, 1985; Lanphere and Frey, 1987; Spengler and Garcia, 1988; Chen and others, 1991; Frey and others, 1994) indicate that the geochemical differences apply not only to the volcanic chain as a whole but to individual volcanoes as well.

The REE reflect the compositional differences and variations in the source region. There is a distinct progression from relatively flat REE abundance patterns and low Ce/Yb ratios in the shield-building volcanics to considerable enrichment in LREE/HREE patterns and high Ce/Yb ratios in the later, more evolved volcanics. Ce/Yb ratios (Fig. 8) are generally less than 25 in the shield-building tholeiites (Roden and others, 1984; Chen and Frey, 1985; Lanphere and Frey, 1987; Tilling and others, 1987; Frey and others, 1994) and mainly greater than 25 in the post-shield and post-erosion volcanics (Clague and Frey, 1982; Chen and Frey, 1985; Lanphere and Frey, 1987; Spengler and Garcia, 1988; Chen and others, 1990; Frey and others, 1990; Frey and others, 1991). Similar behavior also applies to La/Ce ratios (not shown). These results, along with other trace element and isotopic data, are clear indications that the shield-building and later volcanic sequences reflect compositionally distinct mantle sources. Complex models involving mixing of plume, MORB, and wall-rock components, possibly even compositionally distinct diapirs for each volcano, are required to explain the intrashield and intershield volcanic geochemical variations (Griffiths and Campbell, 1991; Eggins, 1992; Frey and Rhodes, 1993; Watson, 1993; Frey and others, 1994; Roden and others, 1994). In general, the shield building stage involves large degrees of melting at relatively shallow mantle depths and the later stages involve lesser melting at greater depths.

9.6.3. ISLAND ARCS

Destructive plate margins are the sites of most of the world's active volcanoes and earthquakes, both of which are associated with the subduction of oceanic lithosphere into the mantle beneath an overriding lithospheric plate. The overriding plate can be either oceanic or continental, in which case the resulting morphological feature of the former is an island arc and the latter is an active continental margin. Both island arcs and active continental margins are characterized by arcuate or linear chains of volcanoes, a deep oceanic trench that parallels the zone of active volcanism on the oceanic side, a dipping zone of seismicity that marks the focus of subduction, and a characteristic igneous association that includes basalt, andesite, dacite, and rhyolite. Because of these characteristics, destructive margins are important in models of lithospheric heterogeneity and formation of continental crust.

The typical island arc basalt to rhyolite association, in which andesitic basalt or andesite is the most abundant type, is generally subdivided on the basis of K_2O abundances

(Peccerillo and Taylor, 1976; Gill, 1981) into an island arc tholeiite suite (low-K), a calc-alkaline suite (medium-K), and a shoshonite suite (high-K). The calc-alkaline suite may be further divided into a normal and a high-K series. Jakes and Gill (1970) were first to recognize that the island arc tholeiite series, characterized by low K and primitive REE patterns, are the dominant rock type in many island arcs.

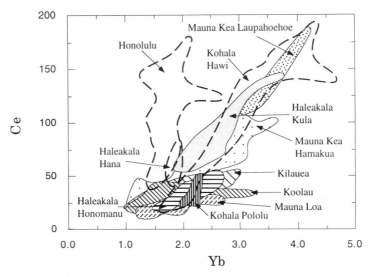

Figure 8: Variations in Ce and Yb abundances in Hawaiian volcanic rocks. Ruled and dashed patterns, shield-building stage (Roden and others, 1984; Chen and Frey, 1985; Lanphere and Frey, 1987; Tilling and others, 1987; Frey and others, 1994); dotted patterns, post-shield stage (Chen and Frey, 1985; Chen and others, 1990; Frey and others, 1990; Frey and others, 1991); transparent pattern, post-erosion stage (Clague and Frey, 1982; Chen and Frey, 1985; Lanphere and Frey, 1987; Spengler and Garcia, 1988; Chen and others, 1990).

The trace element signatures of island arc rocks are distinctive; they are enriched in large ion lithophile elements (LILE) relative to high field strength elements (HFSE) (Hawkesworth and others, 1979a; Kay, 1980; Saunders and others, 1980). A distinguishing feature on chondrite-normalized multi-element variation diagrams (Fig. 9) is a strong depletion in Nb and Ta. This depletion contrasts markedly with trace element contents of divergent boundary (MORB) or within-plate (OIB) basalts that produce relatively smooth concave downward patterns. Inspection of Figure 9 suggests that aside from LILE, Nb, and Ta, island arc rocks are intermediate in composition between OIB and EMORB with respect to concentrations of LREE and

MREE and generally lower in HREE abundances. The trace element patterns of island arc rocks are generally regarded as being derived from slightly enriched mantle source regions containing a residual HFSE phase or from the selected addition of LILE and REE from fluids released by the subducted component along with the concomitant immobilization of the HFSE by the same fluids (Hawkesworth and others, 1993a and references therein).

The REE contents of arc rocks are divisible into two main groups, a low Ce/Yb and high Ce/Yb group (Hawkesworth and others, 1991), with Ce/Yb representing the slope of the REE pattern. Figure 10 shows the slope relationships for a selected group of recent island arcs. The first group has Ce/Yb ratios of about 15 or less and display a general linear relationship between Ce and Yb. The second group has much higher Ce/Yb ratio and an essentially constant Yb content. Most low Ce/Yb arcs have island arc tholeiite or calc-alkaline affinities. Arc rocks with higher Ce/Yb ratios are generally more potassic and typically are high-K calc-alkaline or shoshonite types but may also include low-K varieties. These differences in LREE/HREE ratios are a first order geochemical distinction. They require major differences in the magma source component or in the degree of melting and/or bulk distribution coefficients.

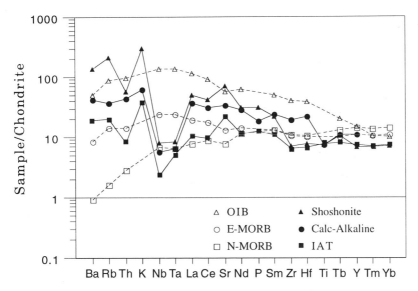

Figure 9: Chondrite normalized multi-element distribution diagram showing distinction among volcanic rocks from the Cretaceous Puerto Rican island arc sequence and average MORB and OIB. Data sources: MORB and OIB (Sun and McDonough, 1989); island arc tholeiite (IAT), average of 8 Rio Majada Group lavas, Puerto Rico (Jolly and others, 1996); calc-alkaline andesite, average of 5 Cerro Gordo Formation lavas, Puerto Rico (unpublished data); shoshonite, average of 6 Perchas Formation lavas, Puerto Rico (unpublished data).

The isotopes of Nd and Sr are also important in helping to elucidate the petrogenesis of island arc volcanic rocks. Figure 11 is a plot of the variation of $^{143}Nd/^{144}Nd$ against $^{87}Sr/^{86}Sr$ of selected arc rocks. The difference between the low and high Ce/Yb trends is also evident on this diagram. The low Ce/Yb arcs have mainly restricted isotopic composition and limited enrichment relative to MORB. The high Ce/Yb arcs show a greater range of isotopic compositions and higher levels of radiogenic isotope enrichment. These characteristics are consistent with the derivation of the high Ce/Yb rocks from source regions with high levels of LREE and other incompatible elements. These features are attributable to the incorporation of subducted sediments in the source melt component and to trace element enriched source regions in the mantle wedge (Hawkesworth and others, 1993a).

Also shown on Figure 11 are vectors directed toward isotopic compositions of hydrothermally altered MORB and toward typical oceanic sediments. Slab-derived fluids from altered MORB have high $^{143}Nd/^{144}Nd$ and high $^{87}Sr/^{86}Sr$, whereas oceanic sediments have low $^{143}Nd/^{144}Nd$ and high $^{87}Sr/^{86}Sr$ (O'Nions and others, 1978; White and others, 1985; Ben Othman and others, 1989). Displacement of some of the arc rocks such as South Sandwich, Tonga-Kermadec, and Grenada toward higher $^{87}Sr/^{86}Sr$ for restricted variation in $^{143}Nd/^{144}Nd$ suggests the intervention of a hydrothermal fluid component. Similarly, the extension of the Aeolian and Philippine arcs to low $^{143}Nd/^{144}Nd$ and high $^{87}Sr/^{86}Sr$ isotopic compositions suggest that at least some of the variation is the result of a sedimentary component from the subducted slab.

9.6.4. CONTINENTAL FLOOD BASALTS

Continental flood basalts, representing major episodes of basaltic magmatism in continental areas, provide an opportunity to evaluate the possible interaction of the continental crust with magmas derived from the sub-continental mantle. Flood basalts typically develop on stable cratons, along continental rift zones, or along continental margins. They represent vast outpourings of lava that is generally erupted in a relatively short time span. A review of Phanerozoic flood basalt provinces is given by Macdougall (1988) and Carlson (1991).

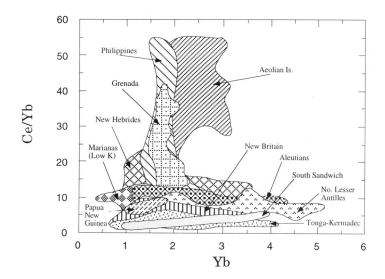

Figure 10: Variation in Ce/Yb ratios of selected recent island arc volcanic rocks having SiO$_2$ ≤ 53.0%.
Data sources: Aeolian Islands (Ellam and others, 1988; Luais, 1988; Ellam and others, 1989;
Francalanci and others, 1993); Aleutians (McCulloch and Perfit, 1981; Miller and others, 1992);
Lesser Antilles (Hawkesworth and others, 1979b; Hawkesworth and Powell, 1980; Baker, 1984;
Thirlwall and Graham, 1984; Davidson, 1986; White and Dupre, 1986; Davidson and others, 1993);
Marianas (Lin and others, 1989; Woodhead, 1989); New Britain (Woodhead and Johnson, 1993); New
Hebrides (Gorton, 1977; Dupuy and others, 1982b; Eggins, 1993); Papua New Guinea (Johnson and
others, 1985; Kennedy and others, 1990); Philippines (Defant and others, 1989; Defant and others,
1990; McDermott and others, 1993b); South Sandwich (Hawkesworth and others, 1977; Cohen and
O'Nions, 1982); Tonga-Kermadec (Ewart and Hawkesworth, 1987).

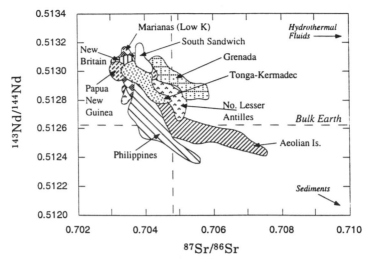

Figure 11: Nd and Sr isotopic ratios of selected recent island arc volcanic rocks having SiO2 ≤ 53.0%. N, average NMORB; E, average EMORB (Sun and McDonough, 1989). Data sources are the same as in Fig. 10

A Precambrian flood basalt province that is preserved in a prominent rift zone is the Keweenawan midcontinent rift system of North America (White, 1960; Green, 1977; Van Schmus, 1992). The main part of the rift system extends from Lake Superior south-southwestward to Nebraska and Kansas, a distance of 1300 km; an eastern arm continues to the southeast into lower Michigan, giving a total length of 2000 km. Vast outpourings of Keweenawan basalt are preserved in the rift. Seismic, potential field, and outcrop data indicate that more than 15 km are preserved at Lake Superior (Cannon, 1989) and a similar thickness is present in the southwestern extent of the rift (Marshall and Lidiak, 1996).

An outstanding feature of the rift system is that it extends through at least four Precambrian basement provinces. These provinces range from the Late Archean (2.7 Ga) greenstone belts, granitic plutonic and gneissic rocks in the Lake Superior region, through a southern region of much older (3.0 to 3.6 Ga) gneissic and migmatitic rocks in southern Minnesota, to the 1.80 to 1.75 Ga granite orthogneisses of the Central Plains Orogen of eastern Nebraska and the 1.67 to 1.66 Ga volcanic and plutonic granitic rocks along the Nebraska-Kansas border. The fact that the rift complex transects diverse Precambrian terranes provides a potential opportunity to evaluate the effects of the interaction of crustal material on the composition of the erupted flood basalts.

The Keweenawan volcanic rocks are divisible into an early and late sequence of lavas (Klewin and Berg, 1991). The early basalts are transitional between tholeiitic and alkalic compositions and have high-Mg basalts as the primitive end member, whereas

the late basalts are mainly tholeiitic and have high-Al olivine tholeiite as the primitive
end member (Berg and Klewin, 1988; Klewin and Shirey, 1992). The early basalts
generally follow a trend of decreasing Mg and increasing Al content; the later basalts
show progressively decreasing Al content. The two main types were apparently derived
from different mantle source regions. These relations are borne out by REE patterns
which show systematic variations with stratigraphic position (Fig. 12). The early lavas
of the Mamainse Point Formation are characterized by enriched LREE relative to
HREE, but also contain lavas having low Ce/Yb ratios. The later Mamainse Point lavas
have low Ce/Yb (<12) ratios, indicative of a relatively depleted mantle source. Similar
low ratios are recorded in lavas of the North Shore Volcanic Group (Ce/Yb < 15), the
Portage Lake Volcanics (Ce/Yb <17), and the Keweenawan basalts of Nebraska (Ce/Yb
< 14). The former two sequences display a distinct linear relationship between Ce and
Yb. There is also a systematic decrease in the LREE/HREE ratio stratigraphically
upward in the North Shore Volcanic Group (Brannon, 1984) and in the Portage Lake
Volcanics (Paces and Bell, 1989), as well as in the Mamainse Point Formation, as
already noted. Similar systematic stratigraphic variation also applies to Nd isotopic
compositions. Both the Mamainse Point and Portage Lake basalt sequences display a
distinct pattern of Nd isotopic evolution from isotopically enriched {negative $\varepsilon_{Nd}(t)$} to
isotopically depleted {positive $\varepsilon_{Nd}(t)$} compositions stratigraphically upward in the
volcanic pile (Paces and Bell, 1989; Klewin and Shirey, 1992). These relations are
clear indication that specific lava sequences underwent a major cycle of chemical
evolution from mantle-enriched to mantle depleted source characteristics.

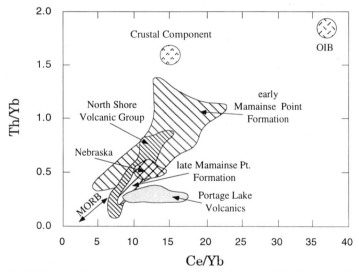

Figure 12: Ce/Yb ratios of Keweenawan basalts. Data sources: Mamainse Point Formation (Klewin
and Berg, 1991); Portage Lake volcanics (Paces and Bell, 1989); North Shore Volcanic Group
(Brannon, 1984); Nebraska (Marshall and Lidiak, 1996).

The REE and isotopic compositional data suggest that a variety of different mantle sources were involved in the generation of Keweenawan magmas (Paces and Bell, 1989; Nicholson and Shirey, 1990; Klewin and Shirey, 1992). Nicholson and Shirey (1990) propose five different source reservoirs: (1) primitive bulk earth mantle; (2) depleted convecting mantle mixed with enriched crust or enriched lithosphere; (3) enriched Archean subcontinental lithospheric mantle; (4) enriched early Proterozoic subcontinental lithospheric mantle; and (5) enriched mantle plume. Most current models invoke two or more distinct mantle sources.

A diagram showing trace element ratios of Keweenawan basalts is shown in Figure 13. These elemental variations suggest that different sources were involved in the generation of the basalts. The earlier, more enriched lavas of each of the sequences lie along broad trends between MORB and a distinct enrichment composition, either an OIB-type component or a crustal/lithospheric component. This latter component could represent either continental crust or early subcontinental lithosphere. Klewin and Shirey (1992) suggest that the lithospheric component could have been island arc crust that was incorporated into the mantle during Archean subduction. The more depleted Keweenawan lavas have compositions that are only slightly more enriched than MORB and they apparently incorporated the enriched lithosphere to a lesser degree.

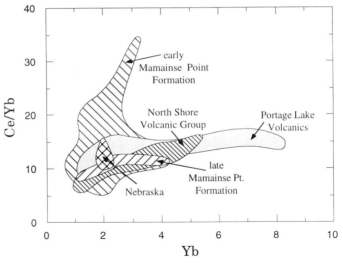

Figure 13: Variation in Ce/Yb vs. Th/Yb in Keweenawan basalts. OIB, ocean island basalt (Sun and McDonough, 1989); crustal component is average bulk continental crust (Taylor and McLennan, 1985). Data sources are the same as in Fig. 12.

9.6.5. GRANITIC ROCKS

Geophysical data indicate that continental lithosphere is about 100-250 km thick and consists of sub-continental mantle overlain by continental crust (Jordan, 1978). The composition of the sub-continental mantle is poorly known, but probably consists of garnet-poor peridotite that is more buoyant than oceanic lithosphere. The overlying continental crust is typically rich in Si, Al, and K and poor in Fe and Mg compared to ocean crust and mantle. It averages about 30 km in thickness and reaches a depth of about 60 km under some high mountain ranges. The average composition of continental crust approaches andesite, with the upper crust (≤ 10km) approximating the composition of granodiorite (Taylor and McLennan, 1985). Granites and related rocks are thus important constituents of upper continental crust.

Much of the early work on the application of REE to the petrogenesis of granitic rocks has been carried out by Hanson and his co-workers (Hanson and Goldich, 1972; Arth and Hanson, 1975; Hanson, 1978, 1980). These studies have shown that trace elements can be extremely useful in determining the origin of granitic rocks.

Next to kimberlites, granitic rocks contain the highest and most variable concentrations of the REE among rock types of general geochemical interest. Interpretation of rare earth patterns in granitic rocks are complicated by a multitude of possible source materials, by different crystallization paths, and by concentrations of REE in accessory minerals as well as in major rock-forming minerals (Cullers and Graf, 1984). Possible modes of origin include: melting and fractionation of sub-continental mantle, subduction and melting of crustal material, melting of crustal material by intrusion and under-plating of large volumes of mafic magma into the crust, and ultrametamorphism and melting of migmatite complexes. These processes tend to concentrate the incompatible elements into the liquid fraction with the result that granites mainly have high concentrations of the REE.

Granitic rocks (quartz diorite, tonalite, granodiorite, granite) exhibit a wide range of REE compositions (Cullers and Graf, 1984), which reflect the chemistry of the source and the crystal/liquid equilibria during crystallization. Plagioclase and/or potassium feldspar are abundant essential minerals in these rocks and many of them have significant Eu anomalies. Europium (divalent state) in these rocks is compatible in the feldspars (Fig. 2), in contrast to the trivalent REE which are incompatible. Removal or concentration of feldspar in a rock can produce either a negative or positive Eu anomaly.

Some REE patterns from representative granitic rocks are summarized in Figure 14. The rocks are divided somewhat arbitrarily into granitoids with negative Eu anomalies and those with positive Eu anomalies (Cullers and Graf, 1984). Both groups are characterized by wide ranges in LREE and HREE contents. Further, the granites in both groups contain larger total REE concentrations than the intermediate rocks.

Among the granitoids with negative Eu anomalies, LREE/HREE ratios that are greater than one are typical. Those with positive anomalies generally have lower LREE/HREE ratios, smaller concentrations of LREE, and a greater range in HREE contents. Melting and crystallization models of plagioclase/melt or potassium feldspar/melt equilibria require that negative Eu anomalies are produced by removal of feldspar from a felsic melt by fractional crystallization or the partial melting of a source rock in which feldspar remains unmelted. Conversely, positive Eu anomalies may represent the concentration of cumulus feldspar in a residue or of feldspar accumulation in a magma. To a lesser extent, hornblende, augite, and hypersthene (Fig. 2) may help produce an Eu anomaly, but in an opposite sense to that of feldspar. Enrichment of the MREE (for example Figure 14b) is controlled mainly by hornblende with which they are compatible.

A measure of the degree of REE fractionation with changing LREE concentration for a series of Phanerozoic granitoid plutonic complexes is shown on Figure 15 as a plot of Ce/Yb against Ce. The granitic rocks are subdivided with minor modification according to the tectonic scheme of Pearce and others (1984) into ocean ridge, island arc, continental margin, continent collision, and anorogenic (within-plate) types. Although the diagram is not intended as a discriminant of tectonic types, several groupings are apparent. N-type plagiogranites and primitive island arc granitoids (Aruba) occupy distinct fields, having relatively low Ce/Yb ratios and low Ce contents. Anorogenic granites are also distinct and are characterized by high levels of LREE (Ce). Conversely, granitoids from mature island arcs, active continental margins, and continental collision zones are indistinguishable from one another in terms of Ce/Yb and Ce concentration. An evaluation of the contrasting REE variations of each of these granitoid complexes from the standpoint of source area, fractional melting, and crystallization history is clearly beyond the scope of the paper. However, the varied geochemical signatures shown by these granitoid suites suggest a variety of magma sources and varying contributions from the subducted slab, mantle wedge, within-plate mantle, and continental crust.

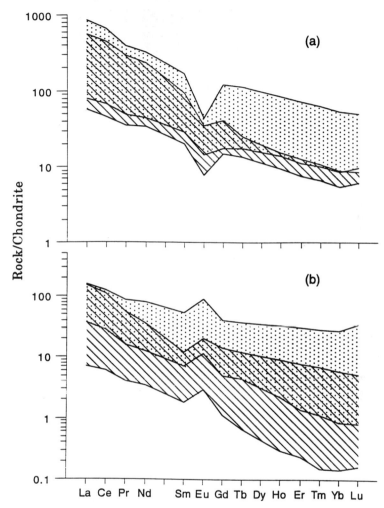

Figure 14: Range of REE compositions in granitic rocks. Ruled pattern, quartz diorite, tonalite, and granodiorite; dot pattern, granite. (a) granitic rocks characterized by negative Eu anomalies. (b) granitic rocks characterized by positive Eu anomalies. Data adapted from Cullers and Graf (1984) and references therein.

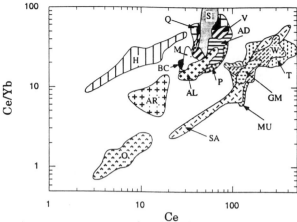

Figure 15: Ce vs. Ce/Yb concentrations in granitoid rocks (SiO$_2$ ≥ 60%) from different tectonic settings. Oceanic areas (inverted v pattern): O, N-type plagiogranite granite (Aldiss, 1981). Island arcs (plus patterns): AR, Aruba batholith (Beets and others, 1984); AL, Aleutians (Perfit and others, 1980). Continental margins (dotted patterns): BC, Coast batholith, British Columbia (Arth and others, 1988); M, Mexico (Bagby and others, 1981); P, Coastal batholith, Peru (Atherton and Sanderson, 1985); S, Sierra Nevada batholith, California (Frey and others, 1978). Continent collision zones (ruled patterns): AD, Adamello massif, Italy (Dupuy and others, 1982a); H, Manaslu, Himalayas (Vidal and others, 1982); V, Vedrette di Ries massif, Italy (Bellieni and others, 1981). Anorogenic granitoids {from data base compiled by Eby (1990)} (dashed patterns): GM, Gabo and Mumbulla, Lachland fold belt, Australia (Collins and others, 1982); MU, Mull (Walsh and others, 1979); SA, Sabaloka complex, Sudan (Harris and others, 1983); T, Trans-Pecos province, Texas (Nelson and others, 1987); W, White Mtn. batholith, New Hampshire (Eby and others, 1992).

9.6.6. SOLUBILITY OF RARE EARTH ELEMENTS

Emphasis so far in this review has been placed on the distribution and behavior of REE in silicate melts using solid/magma partition coefficients. However, it is also important to investigate solubility of these components in aqueous fluids relative to other groups of incompatible elements as many geological processes, including degradation and hydration of high temperature minerals assemblages, take place in the presence of percolating fluids. An efficient measure of the mobility of an element in aqueous fluids is provided by the ionic potential (Z/r), or ratio of ionic charge (Z) to ionic radius of an element in its normal oxidation state (Goldschmidt, 1937; Pearce, 1983). In general, elements with ionic potentials of about three or less are mobile, whereas those with values between about three and seven are immobile. REE and the element Th display intermediate ionic potentials of about 2.5 to 4.5, indicating limited mobility. In contrast, large ion lithophile elements (LILE, including Ba, Rb, K, Pb, and Sr) with Z/r averaging about 1 are highly mobile, and high field strength elements (HFSE, including Nb, Ta, Zr, Hf, and Ti) are highly immobile. The relations are illustrated in Figure 16a, in which selected incompatible elements are listed from left to right in order of decreasing chondrite normalized abundances in NMORB.

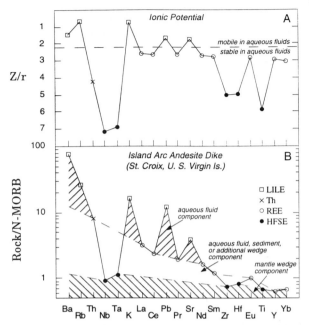

Figure 16: Incompatible element behavior in island arc rocks. A, Ionic potential (Z/r) of some selected incompatible elements, arranged from left to right in order of decreasing incompatibility in NMORB (Sun and McDonough, 1989). B, NMORB-normalized island arc andesite dike (sample SX-8) from St. Croix, U. S. Virgin Islands, illustrating variations in abundances of selected incompatible elements in mantle wedge, aqueous fluid, and sediment plus aqueous fluid components.

Because the REE are normally stable in hydrous geological environments, they are far less likely to undergo redistribution during metamorphic alteration, both in mafic rock (Dickin and Jones, 1983) and in silicate crustal material (Dickin, 1988). In addition, REE remain stable in low temperature environments, such that Nd and its isotopes are resistant to exchange with available fluids. REE patterns generated by magmatic processes are thus retained throughout episodes of extensive recrystallization and other degradational processes. In comparison, Sr and Pb isotopic ratios readily equilibrate with pore fluids and reflect metamorphic or other secondary processes rather than the original magmatic characteristics. For example, during alteration of oceanic crust, Sr isotopes systematically equilibrate with seawater (Jahn and others, 1980), increasing the original mantle ratios considerably, whereas Nd isotopic ratios remain virtually unchanged (Hawkesworth and others, 1993b), even in otherwise completely degraded basalt.

Experimental studies of relative mobility of REE and other incompatible elements in hydrous fluids (Tatsumi and others, 1986) reveal that mobility increases with ionic radius. Solid/aqueous fluid partition coefficients are less well known (Eggler, 1987), but it has long been recognized that the presence of dissolved salts, such as carbonates, hydroxides, and chlorides, promote formation of ionic complexes, thereby significantly increasing solubilities of elements with low ionic potential (Helgeson, 1969). Brennan and Watson (1991) report olivine/fluid partition coefficients (D) for several representative incompatible elements in pure water and concentrated NaCl brines, typical of pore fluids reported from geological environments as diverse as continental basalt sequences, copper porphyry sulphide deposits, and modern island arc volcanic eruptions (Helgeson, 1969). Results reveal that solubilities of LILE in chloride-rich fluids are increased by 1.5 to 3 orders of magnitude, whereas those of REE remain similar to values obtained in pure H_2O (Fig. 17). Thus, two separate sets of partitions coefficients are required to characterize the behavior of these elements in chloride-rich fluids.

These conclusions are particularly applicable to the study of island arc volcanics, as melting in the mantle wedge is generated by introduction of an incompatible element-rich fluid flux from dehydration reactions in the downgoing slab. In general, there is close similarity between ionic potential patterns of selected incompatible elements and patterns produced by island arc volcanic rocks normalized to NMORB (Fig. 16). In both, LREE display slightly higher values than HREE, whereas characteristically LILE are enriched and HFSE are depleted. Incompatible element concentrations in an island arc basalt or andesite therefore can be subdivided into three components. The simplest of these is a component derived exclusively from the mantle wedge source, consisting of material below a line connecting the HFSE (Fig. 16b). The slope of this component varies with the degree of melting and degree of incompatible element enrichment of the wedge source. A second component is material supplied to the melt exclusively by the aqueous fluid flux. This component is represented on Figure 16b by concentrations above a line connecting the relatively immobile REE and Th. Finally, a third or intermediate component of variable origin includes contributions from both sediment (REE) and aqueous fluid (LILE) or, if a decomposing HFSE-bearing phase is present in the source, an additional contribution from the mantle wedge. Because solubility of LILE is several orders of magnitude greater than the solubility of REE, it is commonly suggested (Pearce, 1983; McDermott and others, 1993a) that elevated LREE abundances in island arc volcanics result from melting of source regions that include a LREE-enriched sediment component rather than from addition to the melt of LREE by an aqueous fluid.

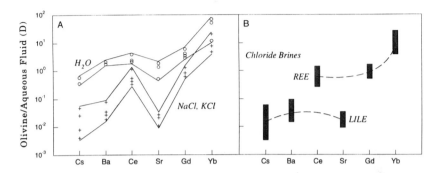

Figure 17: Olivine/aqueous fluid partition coefficients for Cs, Ba, Ce, Sr, Gd, and Yb, arranged in order of incompatibility suggested by Sun and McDonough (1989). A, Comparison of behavior in pure water and in NaCl-rich fluids. B, Partition coefficients (D) of REE and LILE in aqueous fluids.

9.7. Concluding Remarks

This chapter has covered the history and development of the REE in the geological sciences. It is clear from a perusal of the literature that the early geochemists played major roles in developing the geochemical principles that govern the behavior of the REE and in understanding the distribution and abundance of the REE in nature. Much of the recent development of REE research has been tied not only to dedicated researchers but also to vastly improved instrumentation and methodology that is capable of producing reliable results on a routine basis. There has been literally an explosion of REE publications since the 1960s, and the modern history of REE research dates from about that time. The rare earths and other trace elements have been particularly important in helping to establish a geochemical framework to the plate tectonic theory and in placing constraints on geochemical and petrogenetic models of the earth's mantle and crust. Considerable emphasis on REE research continues today.

9.8. References

Ahrens, L. H., 1950, Spectrochemical Analysis: Cambridge, Massachusetts, Addison-Wesley, 269 p.

Ahrens, L. H., 1951, Spectrochemical analysis of some of the rarer elements in the granite and diabase samples: U. S. Geol. Surv. Bull. 980, pt. 4, p. 53-57.

Ahrens, L. H., 1954a, The lognormal distribution of the elements: Geochim. et Cosmochim. Acta, v. 5, p. 49-73.

Ahrens, L. H., 1954b, The lognormal distribution of the elements (2): Geochim. et Cosmochim. Acta, v. 6, p. 121-131.

Ahrens, L. H., 1954c, Quantitative spectrochemical analysis of silicates: London, Pergamon, 122 p.

Ahrens, L. H., 1957, Lognormal-type distributions-III: Geochim. et Cosmochim. Acta, v. 11, p. 205-212.

Ahrens, L. H., and Fleischer, M., 1960, Spectrographic determinations of the major constituents of granite G-1 and diabase W-1: U. S. Geol. Surv. Bull. 1113, p. 83-111.

Aldiss, D. T., 1981, Plagiogranites from the ocean crust and ophiolites: Nature, v. 289, p. 577-578.

Allegre, C. J., and Hart, S. R., eds., 1978, Trace Elements in Igneous Petrology: Dev. Petrology 5, Amsterdam, Elsevier, 272 p.

Allegre, C. J., and Minster, J. F., 1978, Quantitative models of trace element behavior in magmatic processes: Earth Planet. Sci. Lett., v. 38, p. 1-25.

Arth, J. G., Barker, F., and Stern, T. W., 1988, Coast batholith and Taku plutons near Ketchikan, Alaska: petrography, geochronology, geochemistry, and isotopic character: Am. Jour. Sci., v. 288-A, p. 461-489.

Arth, J. G., and Hanson, G. N., 1975, Geochemistry and origin of the early Precambrian crust of northeastern Minnesota: Geochim. et Cosmochim. Acta, v. 39, p. 325-362.

Atherton, M. P., and Sanderson, L. M., 1985, The chemical variation and evolution of the super-units of the segmented Coastal batholith, *in* Picher, W. S., Cobbing, E. J., and Beckinsale, R. S., eds., Magmatism at a Plate Edge, The Peruvian Andes: Glasgow, p. 208-227.

Bagby, W. C., Cameron, K. L., and Cameron, M., 1981, Contrasting evolution of calc-alkalic volcanic and plutonic rocks of western Chihuahua, Mexico: Jour. Geophys. Res., v. 86, p. 10402-10410.

Baker, P. E., 1984, Geochemical evolution of St. Kitts and Montserrat, Lesser Antilles: Geol. Soc. London Journ., v. 141, p. 401-411.

Beets, D. J., Maresch, W. V., Klaver, G. T., Mottana, A., Bocchio, R., Beunk, F. F., and Monen, H. P., 1984, Magmatic rock series and high-pressure metamorphism as constraints on the tectonic history of the southern Caribbean, *in* Bonini, W. E., Hargraves, R. B., and Shagam, R., eds., The Caribbean-South American Plate Boundary and Regional Tectonics: p. 95-130.

Bellieni, G., Peccerillo, A., and Poli, G., 1981, The Vedrette di Ries (Rieserferner) plutonic complex: petrological and geochemical data bearing on its genesis: Contrib. Mineral. Petrol., v. 78, p. 145-156.

Ben Othman, D., White, W. M., and Patchett, J., 1989, The geochemistry of marine sediments, island arc magma genesis, and crust-mantle recycling: Earth Planet. Sci. Lett., v. 94, p. 1-21.

Berg, J. H., and Klewin, K. W., 1988, High-MgO lavas from the Keweenawan midcontinent rift near Mamainse Point, Ontario: Geology, v. 16, p. 1003-1006.

Berman, S., 1957, Determination of yttrium, lanthanum, cerium, neodymium, and ytterbium in test samples granite G-1 and diabase W-1 by a combined chemical-spectrochemical technique: Geochim. et Cosmochim. Acta, v. 12, p. 271-272.

Brannon, J., 1984, Geochemistry of successive lava flows of the Keweenawan North Shore volcanic group [Ph. D. Thesis]: St. Louis, MO, Washington Univ., 312 p.

Brennan, J. M., and Watson, E. B., 1991, Partitioning of trace elements between olivine and aqueous fluids: Earth Planet. Sci. Lett., v. 107, p. 672-688.

Burt, D. M., 1989, Compositional and phase relations among rare earth element minerals, *in* Lipin, B. R., and McKay, G. A., eds., Geochem. Mineral. Rare Earth Elements, Rev. Mineralogy: Washington, D.C., v. 21, Min. Soc. America, p. 259-307.

Cannon, W. F., and others, 1989, The North American midcontinent rift beneath Lake Superior from GLIMPCE seismic reflection profiling: Tectonics, v. 8, p. 305-332.

Carlson, W. R., 1991, Physical and chemical evidence on the cause and source characteristics of flood basalt volcanism: Aust. Jour. Earth Sci., v. 38, p. 525-544.

Chayes, F., 1954, The lognormal distribution of the elements: a discussion: Geochim. et Cosmochim. Acta, v. 6, p. 119-120.

Chen, C.-Y., and Frey, F. A., 1985, Trace element and isotopic geochemistry of lavas from Haleakala volcano, east Maui, Hawaii: implications for the origin of Hawaiian basalts: Jour. Geophys. Res., v. 90, p. 8743-9768.

Chen, C. Y., Frey, F. A., and Garcia, M. O., 1990, Evolution of alkalic lavas at Haleakala volcano, east Maui, Hawaii: Contrib. Mineral. Petrol., v. 105, p. 197-218.

Chen, C. Y., Frey, F. A., Garcia, M. O., Dalrymple, G. B., and Hart, S. R., 1991, The tholeiite to alkalic basalt transition at Haleakala Volcano, Maui, Hawaii: Contrib. Mineral. Petrol., v. 106, p. 183-200.

Clague, D. A., and Dalrymple, G. B., 1987, The Hawaiian-Emperor volcanic chain, part I, geologic evolution: U. S. Geol. Surv. Prof. Paper 1350, p. 5-54.

Clague, D. A., and Frey, F. A., 1982, Petrology and trace element geochemistry of the Honolulu volcanics, Oahu: implications for the oceanic mantle below Hawaii: Jour. Petrology, v. 23, p. 447-504.

Clarke, F. W., 1924, The data of geochemistry: U. S. Geol. Surv. Bull., 841 p.

Clarke, F. W., and Washington, H. S., 1924, The composition of the earth's crust: U. S. Geol. Surv. Prof. Pap. 127, 117 p.

Cohen, R. S., and O'Nions, R. K., 1982, Identification of recycled continental material in the mantle from Sr, Nd, and Pb isotope investigations: Earth Planet. Sci. Lett., v. 61, p. 73-84.

Collins, W. J., Beams, S. D., White, A. J. R., and Chappell, B. W., 1982, Nature and origin of A-type granites with particular reference to southeastern Australia: Contrib. Mineral. Petrol., v. 80, p. 189-200.

Coryell, C. D., Chase, J. W., and Winchester, J. W., 1963, A procedure for geochemical interpretation of terrestrial rare-earth abundance patterns: Jour. Geophys. Res., v. 68, p. 559-566.

Cox, K. G., Bell, J. D., and Pankhurst, R. J., 1979, The Interpretation of Igneous Rocks: London, Allen & Unwin, 450 p.

Cullers, R. L., and Graf, J. L., 1984, Rare earth elements in igneous rocks of the continental crust: intermediate and silicic rocks-ore petrogenesis, *in* Henderson, P., ed., Rare earth element geochemistry: Amsterdam, p. 275-316.

Davidson, J. P., 1986, Isotopic and trace element constraints on the petrogenesis of subduction-related lavas from Martinique, Lesser Antilles: Jour. Geophys. Res., v. 91, p. 5943-5962.

Davidson, J. P., Boghossian, N. D., and Wilson, M., 1993, The geochemistry of the igneous rock suite of St. Martin, northern Lesser Antilles: Jour. Petrology, v. 34, p. 839-866.

Defant, M. J., Jacques, D., Maury, R. C., De Boer, J., and Joron, J. L., 1989, Geochemistry and tectonic setting of the Luzon arc, Philippines: Geol. Soc. America Bull., v. 101, p. 663-672.

Defant, M. J., Maury, R. C., Joron, J. L., Feigenson, M. D., and Leterrier, J., 1990, The geochemistry and tectonic setting of the northern section of the Luzon arc (the Philippines and Taiwan): Tectonophysics, v. 183, p. 187-205.

Dickin, A. P., 1988, Evidence for limited REE leaching from the Roffna Gneiss, Switzerland—a discussion of the paper by Vock *et al.* (1987)(CMP95:145-154): Contrib. Mineral. Petrol., v. 99, p. 273-275.

Dickin, A. P., and Jones, N. W., 1983, Relative mobility during hydrothermal alteration of a basic sill, isle of Skye: Contrib. Mineral. Petrol., v. 82, p. 147-153.

Dupuy, C., Dostal, J., and Fratta, M., 1982a, Geochemistry of the Adamello massif (northern Italy): Contrib. Mineral. Petrol., v. 80, p. 41-48.

Dupuy, C., Dostal, J., Marcelot, G., Bougault, H., Joron, J. L., and Treuil, M., 1982b, Geochemistry of basalts from central and southern New Hebrides arc: implications for their source rock composition: Earth Planet. Sci. Lett., v. 60, p. 207-225.

Eby, G. B., 1990, The A-type granitoids: a review of their occurrence and chemical characteristics and speculations on their petrogenesis: Lithos, v. 26, p. 115-134.

Eby, G. N., Krueger, H. W., and Creasy, J. W., 1992, Geology, geochronology, and geochemistry of the White Mountain batholith, New Hampshire, *in* Puffer, J. H., and Ragland, P. C., eds., Eastern North American Mesozoic Magmatism: Boulder, Geol. Soc. America Spec. Paper 268, p. 379-398.

Eggins, S. M., 1992, Petrogenesis of Hawaiian tholeiites. 2. aspects of dynamic melt segregation: Contrib. Mineral. Petrol., v. 110, p. 398-410.

Eggins, S. M., 1993, Origin and differentiation of picritic arc magmas, Ambae (Aoba), Vanuatu: Contrib. Mineral. Petrol., v. 114, p. 79-100.

Eggler, D. H., 1987, Solubility of major and trace elements in mantle metasomatic fluids: experimental constraints, *in* Menzies, M. A., and Hawkesworth, C. J., eds., Mantle Metasomatism: London, p. 21-41.

Ellam, R. M., Hawkesworth, C. J., Menzies, M. A., and Rogers, N. W., 1989, The volcanism of southern Italy: the role of subduction and the relationship between potassic and sodic alkaline magmatism: Jour. Geophys. Res., v. 94, p. 4589-4601.

Ellam, R. M., Menzies, M. A., Hawkesworth, C. J., Leeman, W. P., Rosi, M., and Serri, G., 1988, The transition from calc-alkaline to potassic orogenic magmatism in the Aeolian Islands, southern Italy: Bull. Volcanol., v. 50, p. 386-398.

Engel, A. E. J., Engel, C. G., and Havens, R. G., 1965, Chemical characteristics of oceanic basalts and the upper mantle: Geol. Soc. America Bull., v. 76, p. 719-734.

Ewart, A. W., and Hawkesworth, C. J., 1987, The Pleistocene to Recent Tonga-Kermadec arc lavas: interpretation of new isotope and rare earth element data in terms of a depleted mantle source model: Jour. Petrology, v. 28, p. 495-530.

Fairbairn, H. W., and others, 1951, A cooperative investigation of precision and accuracy in chemical, spectrochemical and modal analysis of silicate rocks: U. S. Geol. Surv. Bull. 980, 71 p.

Feigenson, M. D., Hofmann, A. W., and Spera, F. J., 1983, Case studies on the origin of basalt II. the transition from tholeiitic to alkalic volcanism on Kohala volcano, Hawaii: Contrib. Mineral. Petrol., v. 84, p. 390-405.

Fersmann, A. E., 1932, Periodic law of abundance of elements: Compt. Rend. Acad. Sci. U. R. S. S., p. 261-266.

Francalanci, L., Taylor, S. R., McCulloch, M. T., and Woodhead, J. D., 1993, Geochemical and isotopic variations in the calc-alkaline rocks of Aeolian arc, southern Tyrrhenian Sea, Italy: constraints on magma genesis: Contrib. Mineral. Petrol., v. 113, p. 300-313.

Frey, F. A., Bryan, W. B., and Thompson, G., 1974, Atlantic Ocean floor: geochemistry and petrology of basalts from Legs 2 and 3 of the Deep Sea Drilling Project: Jour. Geophys. Res., v. 79, p. 5507-5527.

Frey, F. A., Chappell, B. W., and Roy, S. D., 1978, Fractionation of rare-earth elements in the Tuolumne intrusive series, Sierra Nevada batholith, California: Geology, v. 6, p. 239-242.

Frey, F. A., Garcia, M. O., and Roden, M. F., 1994, Geochemical characteristics of Koolau volcano: implications of intershield geochemical differences among Hawaiian volcanoes: Geochim. et Cosmochim. Acta, v. 58, p. 1441-1462.

Frey, F. A., Garcia, M. O., Wise, W. S., Kennedy, A., Gurriet, P., and Albarede, F., 1991, The evolution of Mauna Kea volcano, Hawaii: petrogenesis of tholeiitic and alkalic basalts: Jour. Geophys. Res., v. 96, p. 14,347-14,375.

Frey, F. A., and Haskin, L., 1964, Rare earths in oceanic basalts: Jour. Geophys. Res., v. 69, p. 775-780.

Frey, F. A., and Rhodes, J. M., 1993, Intershield geochemical differences among Hawaiian volcanoes: implications for source compositions, melting process and magma ascent paths: Royal Soc. London Philos. Trans. A, v. 342, p. 121-136.

Frey, F. A., Wise, W. S., Garcia, M. O., West, H., Kwon, S. T., and Kennedy, A., 1990, Evolution of Mauna Kea volcano: petrologic and geochemical constraints on postshield volcanism: Jour. Geophys. Res., v. 95, p. 1271-1300.

Gast, P. W., 1960, Limitations on the composition of the upper mantle: Jour. Geophys. Res., v. 65, p. 1287-1297.

Gast, P. W., 1968, Trace element fractionation and the origin of tholeiitic and alkaline magma types: Geochim. et Cosmochim. Acta, v. 32, p. 1057-1086.

Gill, J. B., 1981, Orogenic Andesites and Plate Tectonics: Berlin, Springer-Verlag, 390 p.

Goldberg, E., Uchiyama, A., and Brown, H., 1951, The distribution of nickel, cobalt, gallium, palladium and gold in iron meteorites: Geochim. et Cosmochim. Acta, v. 2, p. 1-25.

Goldschmidt, V. M., 1937, The principles of distribution of chemical elements in minerals and rocks: Journal of the Chemical Society, v. 140, p. 655-673.

Goldschmidt, V. M., Barth, T., and Lunde, G., 1925, Geochemische Verteilungsgesetze der Elemente. 5. Isomorphie und Polymorphie der Sesquioxyde. Die Lanthaniden-Kontraktion und ihre Konsequenzen: Skrifter utg. av det Norske Videnskaps-akademi i Oslo I. Mat.-naturv. klasse No. 7, 59 p.

Goldschmidt, V. M., Barth, T., Lunde, G., and Zachariasen, 1926, Geochemische Verteilungsgesetze der Elemente. 7. Die Gesetze der Krystallochemie: Skrifter utg. av det Norske Videnskaps-akademi i Oslo 1926 I. Mat.-naturv. klasse No. 2, 117 p.

Goldschmidt, V. M., and Thomassen, L., 1924, Geochemische Verteilungsgesetze der Elemente. 3. Rontgenspektrographische Untersuchungen uber die Verteilung der seltenen Erdmetalle in Mineralen: Vidensk. skrifter. I. Mat.-naturv. klasse No. 5, 58 p.

Goldschmidt, V. M., Ulrich, F., and Barth, T., 1925, Geochemische Verteilungsgesetze der Elemente. 4. Zur Krystallstruktur der Oxyde der seltenen Erdmetalle: Skrifter utg.av det Norske Videnskaps-akademi i Oslo I. Mat.-naturv. klasse No. 5, 24 p.

Gordon, G. E., Randle, K., Goles, G. G., Corliss, J. B., Beeson, M. H., and Oxley, S. S., 1968, Instrumental activation analysis of standard rocks with high-resolution X-ray detectors: Geochim. et Cosmochim. Acta, v. 32, p. 369-396.

Gorton, M. P., 1977, The geochemistry and origin of Quaternary volcanism in the New Hebrides: Geochim. et Cosmochim. Acta, v. 41, p. 1257-1270.

Green, D. H., and Ringwood, A. E., 1967, The genesis of basaltic magmas: Contrib. Mineral. Petrol., v. 15, p. 103-190.

Green, J., and Poldervaart, A., 1955, Some basaltic provinces: Geochim. et Cosmochim. Acta, v. 7, p. 177-188.

Green, J. C., 1977, Keweenawan plateau volcanism in the Lake Superior region, in Baragar, W. R. A., Coleman, L. C., and Hall, J. M., eds., Volcanic Regimes in Canada: Waterloo, Ontario, Geol. Assoc. Canada Spec. Publ. 16, p. 407-422.

Griffiths, R. W., and Campbell, I. H., 1991, On the dynamics of long-lived plume conduits in the convecting mantle: Earth Planet. Sci. Lett., v. 103, p. 214-227.

Hanson, G. N., 1978, The application of trace elements to the petrogenesis of igneous rocks of granitic composition: Earth Planet. Sci. Lett., v. 38, p. 26-43.

Hanson, G. N., 1980, Rare earth elements in petrogenetic studies of igneous systems: Ann. Rev. Earth Planet. Sci., v. 8, p. 371-406.

Hanson, G. N., and Goldich, S. S., 1972, Early Precambrian rocks in the Saganaga Lake-Northern Light Lake area: Geol. Soc. America Mem. 135, p. 179-192.

Harris, N. B. W., Duyverman, H. J., and Almond, D. C., 1983, The trace element and isotope geochemistry of the Sabaloka igneous complex, Sudan: Geol. Soc. London Journ., v. 140, p. 245-256.

Haskin, L., and Gehl, M. A., 1963, The rare-earth contents of standard rocks G-1 and W-1 and their comparison with other rare-earth distribution patterns: Jour. Geophys. Res., v. 68, p. 2037-2043.

Haskin, L. A., 1979, On rare-earth element behavior in igneous rocks, in Ahrens, L. H., ed., Origin and distribution of the elements, Phys. Chem. Earth: Oxford, v. 11, Pergamon, p. 175-189.

Haskin, L. A., and Frey, F. A., 1966, Dispersed and not-so-rare earths: Science, v. 152, p. 299-314.

Haskin, L. A., Haskin, M. A., Frey, F. A., and Wildeman, T. R., 1968, Relative and absolute terrestrial abundances of the rare earths, in Ahrens, L. H., eds., Origin and distribution of the elements: Oxford, p. 889-912.

Haskin, L. A., and Paster, T. P., 1979, Geochemistry and mineralogy of the rare earths, *in* Gschneidner, K. L., Jr., and Eyring, L., eds., Handbook phys. chem. rare earths: Amsterdam, v. 3, North-Holland, p. 1-80.

Hawkesworth, C. J., Gallagher, K., Hergt, J. M., and McDermott, F., 1993a, Mantle and slab contributions in arc magmas: Ann. Rev. Earth Planet. Sci., v. 21, p. 175-207.

Hawkesworth, C. J., Gallagher, K., Hergt, J. M., and McDermott, F., 1993b, Trace element fractionation processes in the generation of island arc basalts: Royal Soc. London Philos. Trans. A, v. 342, p. 179-191.

Hawkesworth, C. J., Hergt, J. M., McDermott, F., and Ellam, R. M., 1991, Destructive margin magmatism and the contributions from the mantle wedge and subducted crust: Aust. Jour. Earth Sci., v. 38, p. 577-594.

Hawkesworth, C. J., Norry, M. J., Roddick, J. C., Baker, P. E., Francis, P. W., and Thorpe, R. S., 1979a, $^{143}Nd/^{144}Nd$, $^{87}Sr/^{86}Sr$, and incompatible element variations in calc-alkaline andesites and plateau lavas from South America: Earth Planet. Sci. Lett., v. 42, p. 45-57.

Hawkesworth, C. J., O'Nions, R. K., and Arculus, R. J., 1979b, Nd and Sr isotope geochemistry of island arc volcanics, Grenada, Lesser Antilles: Earth Planet. Sci. Lett., v. 45, p. 237-248.

Hawkesworth, C. J., O'Nions, R. K., Pankhurst, R. J., Hamilton, P. J., and Evensen, N. M., 1977, A geochemical study of island-arc and back-arc tholeiites from the Scotia sea: Earth Planet. Sci. Lett., v. 36, p. 253-262.

Hawkesworth, C. J., and Powell, M., 1980, Magma genesis in the Lesser Antilles island arc: Earth Planet. Sci. Lett., v. 51, p. 297-308.

Helgeson, H. C., 1969, Thermodynamics of hydrothermal systems at elevated temperatures and pressures: Am. Jour. Sci., v. 267, p. 729-804.

Henderson, P., ed., 1984, Rare Earth Element Geochemistry: Amsterdam, Elsevier, 510 p.

Henderson, P., and Pankhurst, R. J., 1984, Analytical chemistry, *in* Henderson, P., ed., Rare earth element geochemistry: Amsterdam, Elsevier, p. 467-499.

Hess, H. H., 1962, History of ocean basins, *in* Engel, A. E. J., James, H. L., and Leonard, B. F., eds., Petrologic studies: a volume in honor of A. F. Buddington: New York, Geol. Soc. America, p. 599-620.

Hubbard, N. J., 1967, A chemical comparison of ocean ridge, Hawaiian tholeiitic and Hawaiian alkalic basalts: Earth Planet. Sci. Lett., v. 5, p. 346-352.

Humphris, S. E., and Thompson, G., 1983, Geochemistry of rare earth elements in basalts from the Walvis Ridge: implications for its origin and evolution: Earth Planet. Sci. Lett., v. 66, p. 223-242.

Jahn, B., Bernard-Griffiths, J., Charlot, R., Cornichet, J., and Vidal, F., 1980, Nd and Sr isotopic compositions and REE abundances of Cretaceous MORB (Holes 417D and 418A, Legs 51, 52, and 53): Earth Planet. Sci. Lett., v. 48, p. 171-184.

Jakes, P., and Gill, J., 1970, Rare earth elements and the island arc tholeiitic series: Earth Planet. Sci. Lett., v. 9, p. 17-28.

Johnson, R. W., Jaques, A. L., Hickey, R. L., McKee, C. O., and Chappell, B. W., 1985, Manam Island, Papua New Guinea: petrology and geochemistry of a low-TiO_2 basaltic island-arc volcano: Jour. Petrology, v. 26, p. 283-323.

Jolly, W. T., Lidiak, E. G., Dickin, A.P., and Wu, T.-W., 1996, Geochemical diversity of Mesozoic island arc tectonic blocks in eastern Peurto Rico: Geological Society of America Special paper. Accepted for publication..

Jones, G., Sano, H., and Valsami-Jones, E., 1993, Nature and tectonic setting of accreted basalts from the Mino terrane, central Japan: Geol. Soc. London Journ., v. 150, p. 1167-1181.

Jordan, T. H., 1978, Composition and development of the continental tectosphere: Nature, v. 274, p. 544-548.

Kay, R. W., 1980, Volcanic arc magmas: implications of a melting-mixing model for element recycling in the crust-mantle system: Jour. Geology, v. 88, p. 497-522.

Kay, R. W., and Gast, P. W., 1975, The rare earth content and origin of alkali-rich basalts: Jour. Geology, v. 81, p. 653-682.

Kay, R. W., and Hubbard, N. J., 1978, Trace elements in ocean ridge basalts: Earth Planet. Sci. Lett., v. 38, p. 95-116.

Kennedy, A. W., Hart, S. R., and Frey, F. A., 1990, Composition and isotopic constraints on the petrogenesis of alkaline arc lavas: Lihir Island, Papua New Guinea: Jour. Geophys. Res., v. 95, p. 6929-6942.

Klewin, K. W., and Berg, J. H., 1991, Petrology of the Keweenawan Mamainse Point lavas, Ontario: petrogenesis and continental rift evolution: Jour. Geophys. Res., v. 96, p. 457-474.

Klewin, K. W., and Shirey, S. B., 1992, The igneous petrology and magmatic evolution of the Midcontinent rift system: Tectonophysics, v. 213, p. 33-40.

Kuno, H., 1959, Origin of Cenozoic petrographic provinces of Japan and surrounding areas: Bull. Volcanogique, v. 20, p. 37-76.

Lanphere, M. A., and Frey, F. A., 1987, Geochemical evolution of Kohala Volcano, Hawaii: Contrib. Mineral. Petrol., v. 95, p. 100-113.

Le Roex, A. P., 1987, Source regions of mid-ocean ridge basalts: evidence for enrichment processes, *in* Menzies, M. A., and Hawkesworth, C. J., eds., Mantle metasomatism: London, p. 389-422.

Le Roex, A. P., Dick, H. J. B., Erlank, A. J., Reid, A. M., Frey, F. A., and Hart, S. R., 1983, Geochemistry, mineralogy and petrogenesis of lavas erupted along the southwest Indian ridge between the Bouvet triple junction and 11 degrees East: Jour. Petrology, v. 24, p. 267-318.

Lidiak, E. G. and Jolly, W.T., 1996, Geochemistry of instrusive igneous rocks, St. Croix, U.S. Virgin Islands: Geological Society of America Special Paper. Accepted for publication.

Lin, P. N., Stern, R. J., and Bloomer, S. H., 1989, Shoshonitic volcanism in the northern Mariana arc 2. large-ion lithophile and rare earth element abundances: evidence for the source of incompatible element enrichments in intraoceanic arcs: Jour. Geophys. Res., v. 94, p. 4497-4514.

Lipin, B. R., and McKay, G. A., eds., 1989, Geochemistry and Mineralogy of Rare Earth Elements: Rev. Mineralogy, v. 21, Washington, D. C., Min. Soc America, 348 p.

Longerich, H. P., Jenner, G. A., Fryer, B. J., and Jackson, S. E., 1990, Inductively coupled plasma-mass spectrometric analysis of geological samples: a critical evaluation based on case studies: Chemical Geol., v. 83, p. 105-118.

Luais, B., 1988, Mantle mixing and crustal contamination as the origin of the high-Sr radiogenic magmatism of Stromboli (Aeolian arc): Earth Planet. Sci. Lett., v. 88, p. 93-106.

Macdonald, G. A., and Katsura, T., 1964, Chemical composition of Hawaiian lavas: Jour. Petrology, v. 5, p. 82-133.

Macdougall, J. D., ed., 1988, Continental Flood Basalts: Dordrecht, Kluwer, 341 p.

Marshall, L. P., and Lidiak, E. G., 1996, Geochemistry and paleomagnetism of Keweenawan basalts in the subsurface of Nebraska: Precambrian Res., v. 76, p. 47-65.

Mason, B., 1992, Victor Moritz Goldschmidt: Father of Modern Geochemistry: San Antonio, TX, Geochem. Soc., Spec. Publ. no. 4, 184 p.

Masuda, A., 1962, Regularities in variation of relative abundances of lanthanide elements and an attempt to analyze separation-index patterns of some minerals: Journal of Earth Science Nagoya University, v. 10, p. 173-187.

McCulloch, M. T., and Perfit, M. R., 1981, 143Nd/144Nd, 87Sr/86Sr and trace element constraints on the petrogenesis of Aleutian island arc magmas: Earth Planet. Sci. Lett., v. 56, p. 167-179.

McDermott, F., Defant, M. J., Hawkesworth, C. J., Maury, R. C., and Joron, J. L., 1993a, Isotope and trace element evidence for three component mixing in the genesis of the North Luzon arc lavas (Philippines): Contrib. Mineral. Petrol., v. 113, p. 9-23.

McDermott, F., Defant, M. J., Hawkesworth, C. J., Maury, R. C., and Joron, J. L., 1993b, Isotope and trace element evidence for three component mixing in the genesis of the North Luzon arc lavas (Philippines): Contrib. Mineral. Petrol., v. 113, p. 9-23.

McKay, G. A., 1989, Partitioning of rare earth elements between major silicate minerals and basaltic melts, *in* Lipin, B. R., and McKay, G. A., eds., Geochem. Mincral. Rarc Earth Elements, Rev. Mineralogy: Washington, D. C., v. 21, Min. Soc. America, p. 45-77.

Miller, D. M., Langmuir, C. H., Goldstein, S. L., and Franks, A. L., 1992, The importance of parental magma composition to calc-alkaline and tholeiitic evolution: evidence from Umnak Island in the Aleutians: Jour. Geophys. Res., v. 97, p. 321-343.

Miller, R. L., and Goldberg, E. D., 1955, The normal distribution in geochemistry: Geochim. et Cosmochim. Acta, v. 8, p. 53-62.

Minami, E., 1935, Gehalte an seltenen Erden in europaischen und jananischen Tonschiefern: Nachr. Ges. Wiss. Gottingen, Z, Math-Physik KL. IV, v. 1, p. 155-170.

Mitchell, R. L., 1938, The spectrographic determination of trace elements in soils. I. the cathode layer arc: Journal of the Society of Chemical Industry Transactions, v. 57, p. 210-213.

Mitchell, R. L., 1948, Spectrographic analysis of soils, plants, and related materials: Harpenden, England, Commonwealth Bur. Soil Sci., Tech. Comm. 44, 183 p.

Moeller, T., 1963, The Chemistry of the Lanthanides: New York, Reinhold, 117 p.

Nelson, D. O., Nelson, K. L., Reeves, K. D., and Mattison, G. D., 1987, Geochemistry of tertiary alkaline rocks of the eastern Trans-Pecos magmatic province, Texas: Contrib. Mineral. Petrol., v. 97, p. 72-92.

Neuman, H., Mead, J., and Vitaliano, C. J., 1954, Trace element variation during fractional crystallization as calculated from the distribution law: Geochim. et Cosmochim. Acta, v. 6, p. 90-99.

Nicholson, S. W., and Shirey, S. B., 1990, Midcontinent rift volcanism in the Lake Superior Region: Sr, Nd, and Pb isotopic evidence for a mantle plume origin: Jour. Geophys. Res., v. 95, p. 10851-10868.

Nockolds, S. R., and Allen, R., 1953, The geochemistry of some igneous rocks: Geochim. et Cosmochim. Acta, v. 4, p. 105-142.

Nockolds, S. R., and Allen, R., 1954, The geochemistry of some igneous rocks: Part II: Geochim. et Cosmochim. Acta, v. 5, p. 245-285.

Nockolds, S. R., and Allen, R., 1956, The geochemistry of some igneous rocks-III: Geochim. et Cosmochim. Acta, v. 9, p. 34-77.

O'Hara, M. J., 1965, Primary magmas and the origin of basalts: Scot. Jour. Geol., v. 1, p. 19-40.

O'Nions, R. K., Carter, S. R., Cohen, R. S., Evensen, N. M., and Hamilton, P. J., 1978, Nd and Sr isotopes in oceanic ferro-manganese deposits and ocean floor deposits: Nature, v. 273, p. 435-438.

Paces, J. B., and Bell, K., 1989, Non-depleted sub-continental mantle beneath the Superior Province of the Canadian shield: Nd-Sr isotopic and trace element evidence from midcontinent rift basalts: Geochim. et Cosmochim. Acta, v. 53, p. 2023-2035.

Pearce, J. A., 1983, Role of the sub-continental lithosphere in magma genesis at active continental margins, *in* Hawkesworth, C. J., and Norry, M. J., eds., Continental basalts and mantle xenoliths: Cheshire, U. K., p. 230-249.

Pearce, J. A., Harris, N. B. W., and Tindle, A. G., 1984, Trace element discrimination diagrams for the tectonic interpretation of granitic rocks: Jour. Petrology, v. 25, p. 956-983.

Peccerillo, A., and Taylor, S. R., 1976, Geochemistry of Eocene calc-alkaline volcanic rocks from the Kastamonu area, northern Turkey: Contrib. Mineral. Petrol., v. 58, p. 63-91.

Perfit, M. R., Brueckner, H., Lawrence, J. R., and Kay, R. W., 1980, Trace element and isotopic variations in a zoned pluton and associated volcanic rocks, Unalaska Island, Alaska: a model for fractionation in the Aleutian calcalkaline suite: Contrib. Mineral. Petrol., v. 73, p. 69-87.

Rankama, K., and Sahama, T. G., 1950, Geochemistry: Chicago, Univ. Chicago, 912 p.

Roden, M. F., Frey, F. A., and Clague, D. A., 1984, Geochemistry of tholeiitic and alkalic lavas from the Koolau Range, Oahu, Hawaii: implications for Hawaiian volcanism: Earth Planet. Sci. Lett., v. 69, p. 141-158.

Roden, M. F., Trull, T., Hart, S. R., and Frey, F. A., 1994, New He, Nd, Pb, and Sr isotopic constraints on the constitution of the Hawaiian plume: results from Koolau volcano, Oahu, Hawaii: Geochim. et Cosmochim. Acta, v. 58, p. 1437-1446.

Rollin, E. S., and others, 1960, Second report on a cooperative investigation of the composition of two silicate rocks: U. S. Geol. Surv. Bull. 1113, 126 p.

Rollinson, H., 1993, Using Geochemical Data: Evaluation, Presentation, Interpretation: Essex, England, Longman, 352 p.

Saunders, A. D., Tarney, J., and Weaver, S. D., 1980, Transverse geochemical variations across the Antarctic Peninsula: implications for the genesis of calc-alkaline magmas: Earth Planet. Sci. Lett., v. 46, p. 344-360.

Schilling, J. G., and Winchester, J. W., 1966, Rare earths in Hawaiian basalts: Science, v. 153, p. 867-869.

Schilling, J. G., and Winchester, J. W., 1969, Rare-earth contribution to the origin of Hawaiian lavas: Contrib. Mineral. Petrol., v. 23, p. 27-37.

Schilling, J. G., Zajac, M., Evans, R., Johnston, T., White, W., Devine, J. D., and Kingsley, R., 1983, Petrologic and geochemical variations along the mid-Atlantic ridge from 29°N to 73°N: Am. Jour. Sci., v. 283, p. 510-586.

Schmitt, R. A., Smith, R. H., Lasch, J. E., Mosen, A. W., Olchy, D. A., and Vasilevskis, J., 1963, Abundances of the fourteen rare-earth elements, scandium and yttrium in meteorites and terrestrial matter: Geochim. et Cosmochim. Acta, v. 27, p. 577-622.

Schnetzler, C. C., and Philpotts, J. A., 1970, Partition coefficients of rare-earth elements between igneous matrix material and rock-forming mineral phenocrysts-II: Geochim. et Cosmochim. Acta, v. 34, p. 331-340.

Shannon, R. D., 1976, Revised effective ionic radii and systematic studies of interatomic distances in halides and chalcogenides: Acta Crystallographica, v. A32, p. 751-767.

Shannon, R. D., and Prewitt, C. T., 1970, Revised values of effective ionic radii: Acta Crystallographica, v. B26, p. 1046-1048.

Shaw, D. M., 1970, Trace element fractionation during anatexis: Geochim. et Cosmochim. Acta, v. 34, p. 237-243.

Smales, A. A., Mapper, D., and Wood, A. J., 1957, The determination, by radioactivation, of small quantities of nickel, cobalt and copper in rocks, marine sediments and meteorites: Analyst, v. 82, p. 75-88.

Smith, D. M., and Wiggins, G. M., 1949, Analysis of rare earth oxides by means of emission spectra: Analyst, v. 74, p. 95-101.

Spengler, S. R., and Garcia, M. O., 1988, Geochemistry of the Hawi lavas, Kohala volcano, Hawaii: Contrib. Mineral. Petrol., v. 99, p. 90-104.

Strock, L. W., 1936, Spectrum Analysis with the Carbon Arc Cathode Layer (Glimmschicht): London, Adam Hilger, 54 p.

Sun, S.-s., and Hanson, G. N., 1975, Origin of Ross Island basanitoids and limitations upon the heterogeneity of mantle sources for alkali basalts and nephelinites: Contrib. Mineral. Petrol., v. 52, p. 77-106.

Sun, S.-s., and McDonough, W. F., 1989, Chemical and isotopic systematics of oceanic basalts: implications for mantle composition and processes, *in* Saunders, A. D., and Norry, M. J., eds., Magmatism in Ocean Basins: Cheshire, England, p. 267-305.

Sun, S.-s., Nesbitt, R. W., and Sharaskin, A. Y., 1979, Geochemical characteristics of mid-ocean ridge basalts: Earth Planet. Sci. Lett., v. 44, p. 119-138.

Tatsumi, Y., Hamilton, D. L., and Nesbitt, R. W., 1986, Chemical characteristics of fluid phase released from a subducted lithosphere and origin of arc magmas: evidence from high-pressure experiments and natural rocks: Jour. Volcan. Geotherm. Res., v. 29, p. 293-309.

Taylor, S. R., and McLennan, S. M., 1985, The Continental Crust: its Composition and Evolution: Oxford, Blackwell, 312 p.

Thirlwall, M. F., 1982, A triple-filament method for rapid and precise analysis of rare-earth elements by isotope dilution: Chemical Geol., v. 35, p. 155-166.

Thirlwall, M. F., and Graham, A. M., 1984, Evolution of high-Ca, high Sr C-series basalts from Grenada, Lesser Antilles: the effects of intra-crustal contamination: Geol. Soc. London Journ., v. 141, p. 427-445.

Tilling, R. I., Wright, T. L., and Millard, H. T., Jr., 1987, Trace-element chemistry of Kilauea and Mauna Loa lava in space and time: a reconnaissance: U. S. Geol. Surv. Prof. Pap. 1350, p. 641-689.

Topp, N. E., 1965, The Chemistry of the Rare-Earth Elements: Amsterdam, Elsevier, 164 p.

Towell, D. G., Winchester, J. W., and Volfovsky Spirn, R., 1965, Rare-earth distributions in some rocks and associated minerals of the batholith of southern California: Jour. Geophys. Res., v. 70, p. 3485-3496.

Van Schmus, W. R., 1992, Tectonic setting of the midcontinent rift system: Tectonophysics, v. 213, p. 1-15.

Vickery, R. C., 1953, Chemistry of the Lanthanons: New York, Academic Press, 296 p.

Vickery, R. C., 1961, Analytical Chemistry of the Rare Earths, in Belcher, R., and Gordon, L., eds., Analytical Chemistry: Oxford, Pergamon Press, v. 3, 139 p.

Vidal, P., Cocherie, A., and Le Fort, P., 1982, Geochemical investigations of the origin of the Manaslu leucogranite (Himalaya, Nepal): Geochim. et Cosmochim. Acta, v. 46, p. 2279-2292.

Vine, F. J., and Matthews, D. H., 1963, Magnetic anomalies over oceanic ridges: Nature, v. 199, p. 947-949.

Wager, L. R., and Mitchell, R. L., 1951, The distribution of trace elements during strong fractionation of basic magma—a further study of the Skaergaard intrusion, east Greenland: Geochim. et Cosmochim. Acta, v. 1, p. 129-208.

Wager, L. R., and Mitchell, R. L., 1953, Trace elements in a suite of Hawaiian lavas: Geochim. et Cosmochim. Acta, v. 3, p. 217-223.

Walsh, J. N., Beckinsale, R. D., Skelhorn, R. R., and Thorpe, R. S., 1979, Geochemistry and petrogenesis of Tertiary granitic rocks from the island of Mull, northwest Scotland: Contrib. Mineral. Petrol., v. 71, p. 99-116.

Watson, S., 1993, Rare earth element inversions and percolation models for Hawaii: Jour. Petrology, v. 34, p. 763-783.

White, W. M., and Dupre, B., 1986, Sediment subduction and magma genesis in the Lesser Antilles: isotopic and trace element constraints: Jour. Geophys. Res., v. 91, p. 5927-5941.

White, W. M., Dupre, B., and Vidal, P., 1985, Isotope and trace element geochemistry of sediments from the Barbados ridge-Demerara plain region, Atlantic Ocean: Geochim. et Cosmochim. Acta, v. 49, p. 1875-1886.

White, W. S., 1960, The Keweenawan lavas of Lake Superior: an example of flood basalts: Am. Jour. Sci., v. 285A, p. 367-374.

Wood, B. J., and Fraser, D. G., 1978, Elementary Thermodynamics for Geologists: Oxford, Oxford Univ. Press, 303 p.

Wood, D. A., Joron, J. L., Treuil, M., Norry, M., and Tarney, J., 1979, Elemental and Sr isotope variations in basic lavas from Iceland and the surrounding ocean floor: Contrib. Mineral. Petrol., v. 70, p. 319-339.

Woodhead, J. D., 1989, Geochemistry of the Mariana arc (western Pacific): source composition and processes: Chemical Geol., v. 76, p. 1-24.

Woodhead, J. D., and Johnson, R. W., 1993, Isotopic and trace-element profiles across the New Britain island arc, Papua New Guinea: Contrib. Mineral. Petrol., v. 113, p. 479-491.

Yoder, H. S., and Tilley, C. E., 1962, Origin of basalt magmas: an experimental study of natural and synthetic rock systems: Jour. Petrology, v. 3, p. 342-532.

CHAPTER 10

USE OF LANTHANUM AS A TOOL TO DELINEATE CALCIUM
MOBILIZATION PATTERNS IN SMOOTH MUSCLE

GEORGE B. WEISS
M. Hurley & Associates, Inc.
571 Central Avenue
Murray Hill, NJ 07974

10.1. Introduction

The lanthanum-related literature in physiology and pharmacology has become both
extensive and diverse during the preceding two decades. Use of lanthanum and similar
lanthanides as tools to elucidate various actions and pathways of calcium has resulted in
advances in our understanding of many important roles of calcium ion in such
fundamental processes as excitation-contraction coupling and stimulus-secretion
coupling in a wide variety of cellular systems. Dissection with lanthanides of various
calcium components present in complex biological preparations has also facilitated
identification of more precise cellular mechanisms of drug action and led to more
rational approaches for delineation of activities of important therapeutic agents.

In this chapter, I will not attempt to summarize this active field of research. Instead, I
will develop a narrative sequence based primarily upon my own personal research
activities in order to illustrate both the origins of concepts and the accumulation of
cellular information in this area of lanthanide-related research. Initially, I will discuss
the use of lanthanum to help dissociate calcium mobilization patterns in smooth muscle
and to define the effects and spectrum of activity of drugs on specific calcium
components in different types of smooth muscle. Subsequently, extending this to other
systems, I will attempt to summarize the models that were developed.

10.2. The Research Problem and the Idea to Use Lanthanum

I first became interested in smooth muscle and calcium during my graduate studies at
Vanderbilt University with Dr. Leon Hurwitz when I examined the differing calcium-
dependent actions of ethanol and cocaine (1,2). The complexity of the interactions
clearly indicated that interference with one or more cellular action of calcium were

C. H. Evans (ed.), Episodes from the History of the Rare Earth Elements, 189–203.
© 1996 Kluwer Academic Publishers. Printed in the Netherlands.

involved in the actions of these two agents. At that time, Drs. A.M. Shanes and C.P. Bianchi were doing their initial studies on ^{45}Ca fluxes in striated muscles (e.g., 3), and I went to their laboratory at the University of Pennsylvania as a Postdoctoral Fellow to learn these techniques so that I could apply them to smooth muscle preparations. Unfortunately, the ^{45}Ca uptake and efflux procedures that so successfully separated components in the structurally developed striated muscles were not as readily resolved in the less anatomically differentiated smooth muscle cellular preparations.

Lacking a clear way to dissociate calcium components, I pursued other research directions initially after my move to the Medical College of Virginia including such related studies in smooth muscle as the precise measurement of extracellular space and tissue compartments. However, in 1964, a paper by Lettvin and co-workers appeared (4) that turned my attention toward the use of lanthanum ion (La^{+++}). They predicted that La^{+++}, with an ionic radius similar to Ca^{++} and a higher valence, would bind at superficially located Ca^{++} sites in a less reversible manner than does Ca^{++}. Subsequently (5), they demonstrated that La^{+++} exerts a blocking action in lobster axons similar to that of an extremely high Ca^{++} concentration. The idea that La^{+++} was active at extracellular sites but did not penetrate into the cell made its potential as a tool in smooth muscle particularly attractive because evidence had accumulated that attempts to measure specific physiological and pharmacological alterations in Ca^{++} distribution and movements (as ^{45}Ca fluxes) were unsuccessful. The major portion of Ca^{++} in smooth muscle (approximately 80-90%) was not within the cell and contributed to quantitatively large and unrelated ion movements that obscured the relatively smaller specific Ca^{++} changes of interest. Thus, use of La^{+++} as a tool to block some but not all Ca^{++} movements in isolated smooth muscle preparations appeared to be a promising experimental approach.

At that time, Frank Goodman had joined my laboratory as a predoctoral student, and I suggested that he do the experiments with La^{+++}, possibly as an eventual Doctoral Thesis. The initial experimental studies we performed with La^{+++} were, first, in guinea pig intestinal (ileal) smooth muscle (6) and, subsequently, in rat uterine smooth muscle (7) and rabbit aortic smooth muscle (8). The results obtained confirmed that La^{+++} appeared to replace and/or displace the large Ca^{++} components at extracellular and superficial sites and to prevent uptake of Ca^{++} to various cellular sites. Differences among the types of smooth muscle used were also noted both in patterns of Ca^{++} accumulation and in responsiveness to stimulatory agents. The amount of Ca^{++} bound (and displaced by La^{+++}) was much greater in vascular (rabbit aortic) smooth muscle (8) than in ileal (6) or uterine (7) smooth muscle. Also, the sensitivity to La^{+++} of K^+-induced responses was greater than that of responses to stimulatory agents in aortic (8) or uterine (7) preparations but not in ileal ones (6). Thus, use of La^{+++} could block specific calcium components involved in drug actions.

10.3. The Lanthanum Method

At this point, the ability to quantitatively delineate the calcium components present in smooth muscle was greatly improved by an approach developed by Dr. C. van Breemen and associates (9). This method, called the lanthanum method, was based upon the assumptions that a sufficiently high concentration of extracellular La^{+++} will (a) displace and replace extracellular Ca, (b) block both Ca^{++} uptake and efflux, and (c) not enter the cell in significant quantities to affect cellular Ca^{++} distribution. The procedure used exposed tissues to a variety of stimulatory agents or conditions in the presence of ^{45}Ca and subsequently placed them into washout solutions containing a concentration of La^{+++} high enough to replace all extracellular or superficial Ca^{++} and to prevent any further uptake or efflux of cellular ^{45}Ca. In this manner, effects on cellular ^{45}Ca uptake obscured by the much larger quantities of extracellular ^{45}Ca present and by nonspecific ^{45}Ca movements would be detected. With these procedures, they were able to identify correlations between contractile responses in aortic smooth muscle and mobilization of specific Ca^{++} stores (10).

Though a number of technical modifications of our experimental procedures were subsequently developed to validate the assumptions in the lanthanum method in different preparations (and these will be described in several of the following sections), the basic approach was an extremely useful and productive one. Application of the lanthanum method to different types of smooth muscles led to comparative quantitative measurements that contributed to comprehensive explanations of variations in smooth muscle physiology and responsiveness to various pharmacological agents.

10.4. Validation and Use of the Lanthanum Method

Of the three major assumptions of the lanthanum method, the ones involving blockade of Ca^{++} uptake and displacement of extracellular and superficial Ca^{++} were easiest to validate. Earlier studies had shown extensive displacement of Ca^{++}, but the critical measurement involved increasing the La^{+++} concentration in the bathing solution during exposure to ^{45}Ca so that the uptake of ^{45}Ca did not exceed the uptake of ^{14}C-labeled sugars that did not penetrate through or bind to the cell membrane and, therefore, could be used to estimate the size of the extracellular compartment. This type of measurement was first reported by Burton and Godfraind (11) in guinea pig ileal longitudinal smooth muscle, and similar measurements have subsequently been obtained in numerous other smooth muscle preparations.

The relationship between La^{+++} and efflux of Ca^{++} was a more difficult problem. Efflux of ions is composed of a number of processes (including passive diffusion, ion exchange, and active transport mechanisms), many of which are incompletely characterized. In an initial study, Deth (12) found that lowering the temperature of the solution bathing rabbit aortae from $37°$ to $2°C$ during a 60 min washout of ^{45}Ca doubled the amount of cellular ^{45}Ca retained. Also, Brading and Widdicombe reported (13) that La^{+++} altered monovalent ion movements in smooth muscle. In an attempt to minimize

^{45}Ca efflux in the presence of La^{+++}, Dr. H. Karaki (who was on a sabbatical visit from the University of Tokyo to my laboratory at the University of Texas Southwestern Medical School in Dallas) and I developed a solution and conditions under which efflux of ^{45}Ca was markedly decreased (14). Use of this solution, in which La^{+++} was substituted for all other cations (to 80.8 mM) and the bathing solution temperature lowered to 0.5°C, during ^{45}Ca washout further increased the $t_{1/2}$ of washout of the slower washout component in rabbit aortic smooth muscle from 136 min (in 10 mM added La^{+++} solution at 34°C) to approximately 630 min (Fig. 1). At this slower rate, the loss of cellular ^{45}Ca during a 60 min washout is very small. Under these conditions, the retention of Ca^{++} (expressed as residual Ca^{++} uptake or as Ca^{++} retention after La^{+++}) taken up previously as ^{45}Ca in the presence and absence of depolarizing concentrations of norepinephrine (NE) and K^{+} could be measured with increased quantitative precision (Table 1). Furthermore, it was possible to measure rapid changes in Ca^{++} uptake (15) in sets of muscles obtained from the same animal in which differently-treated muscles were removed from incubation solutions, washed out in La^{+++} solution, and analyzed for ^{45}Ca (Fig. 2).

The question of whether the cell membrane was permeable to lanthanides was addressed in a collaborative study with Dr. F. Goodman in which the distribution of the radioactive lanthanide, promethium (^{147}Pm) was measured. In this study (16), it was found that significant quantities of ^{147}Pm do not appear to be accumulated within the cell and that distribution of ^{147}Pm could be most simply described in terms of binding at and desorption from surface-accessible fiber sites.

10.5. Effects of Other Rare Earth Ions

The success of experimental approaches with La^{+++} suggested to Dr. Goodman and me that some of the other rare earth ions might differ significantly from La^{+++} in their degree of access to various Ca^{++} sites or stores, and they might be of additional value in further delineation of important cellular actions of Ca^{++}. Thus, to determine whether Lu^{+++}, Eu^{+++}, and Nd^{+++} interacted with Ca^{++} in a manner similar to that of La^{+++}, the activities of these ions were characterized in rabbit aortic smooth muscle preparations (17). It was found that the actions of Lu^{+++} were quantitatively similar to those of approximately 20-fold lower La^{+++} concentrations, whereas those of Nd^{+++} and Eu^{+++} resembled effects obtained with equimolar concentrations of La^{+++}. Thus, the differences observed primarily with equimolar Lu^{+++} could be attributed solely to a decreased affinity for superficial Ca^{++} binding sites and a decreased ability to block Ca^{++} (^{45}Ca) uptake.

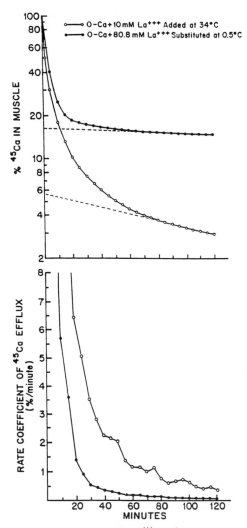

Figure 1: Comparison of effects of added (10 mM) La[+++] at 34°C with those of substituted (80.8 mM) La[+++] at 0.5°C on [45]Ca efflux from rabbit aortic smooth muscle into O-Ca solution. Desaturation (top) and corresponding rate coefficient (bottom) plots were obtained from averaged washouts of the same muscles. Dashed lines indicate extrapolation of the slow washout components to the beginning of the washout. All muscles were incubated in normal Ca[++] solutions plus [45]Ca for 60 min before the washout. From (14)

Table 1 Effects of Ca^{++} Channel Blockers and Nitroprusside on Low Affinity Ca^{++} Retention

^{45}Ca Incubation Conditions	Ca^{++} Retention After La^{+++} (nmol/g ± S.E.)	% Change in Ca^{++} Retention
Control	302 ± 11	------
+ D-600	294 ± 33	-2.8
+ Norepinephrine	299 ± 15	-0.8
+ K$^+$	605 ± 27	+100.4
+ K$^+$ + D-600	299 ± 9	-1.0
+ K$^+$ + Norepinephrine	785 ± 33	+160.0

Adapted from (14).

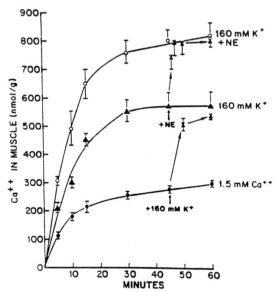

Figure 2: Effects of 160 mM K$^+$ solution and of NE (10^{-6}M) on residual La^{+++}-resistant Ca^{++} uptake from solutions containing ^{45}Ca plus 1.5 mM added Ca^{++}. Muscles were incubated with ^{45}Ca in 1.5 mM Ca^{++}. Tris-buffered solution (•), 160 mM K$^+$ solution in which K$^+$ was isosmotically substituted for Na$^+$ (▲) or 160 mM K$^+$ solution containing 10^{-6} M NE (o). As specified, either muscles were transferred to 160 mM K$^+$ solution containing ^{45}Ca or NE was added to the ^{45}Ca incubation solution at the points indicated by arrows for the durations terminating at (X). All muscles were washed out for 60 min in La^{+++} solution preceding ^{45}Ca measurement. From (15)

The weaker but similar action of Lu^{+++} proved to be useful in a comparison between the actions of this ion on Ca^{++} distribution and movements and those of Sr^{++}, a weak Ca^{++} analog (18). In contrast to Lu^{+++}, Sr^{++} appears to substitute for Ca^{++} in some but not all Ca^{++}-dependent actions in vascular smooth muscle (19). The effects of the two ions on ^{45}Ca desaturation curve components and on ^{45}Ca uptakes plotted on Scatchard-type coordinates were compared. These two techniques were helpful in dissociating fast and slow ^{45}Ca washout components and high and low affinity ^{45}Ca binding sites, respectively. More precise mechanisms of action could also be delineated for Sr^{++} and Lu^{+++}. Strontium ion has a weaker affinity for Ca^{++} binding sites than does Ca^{++} and, consequently, high concentrations of Sr^{++} can displace ^{45}Ca from both extracellular and cellular high affinity sites but not from low affinity sites, whereas Lu^{+++} inhibits ^{45}Ca binding indirectly at both high and low affinity sites by blocking uptake to these sites.

10.6. Studies in Other Systems

Studies with La^{+++} were conducted in other systems primarily to understand what variations in Ca^{++} binding occurred and how these differences affected responsiveness to pharmacological agents. The basis for this was a general hypothesis developed earlier (19) that both the qualitative and quantitative nature of the muscle response to a given agent is directly determined by the amounts of intracellular Ca^{++} available and this, in turn, is a direct consequence of differing Ca^{++}-binding properties of the specific cytoarchitecture present in each particular type of muscle. Thus, the nature of the final response elicited by the drug-receptor interaction is not a function of the receptor activation but, rather, a result of alterations induced by the drug-receptor interaction in membrane binding and permeability characteristics and, subsequently, in the intracellular Ca^{++} level.

Among the studies that we conducted with La^{+++} was a comparison of Ca^{++} binding in renal arteries and veins done in collaboration with Dr. R. Kelly Hester (20). It was found that the two types of vessel were similarly responsive to the stimulatory agents used but that renal arteries bound more Ca^{++} than did veins, and that the Ca^{++} mobilized in arteries was predominantly from these binding sites, whereas the contractile responses induced in veins were more dependent upon uptake of extracellular Ca^{++} at equivalent concentrations. In a study in collaboration with Drs. Goodman, Karaki and Nakagawa (21), guinea pig tracheal smooth muscle was found to have Ca^{++} components similar to those in vascular smooth muscle. However, the components were not as readily dissociated, and the lanthanum method did not slow down Ca^{++} loss sufficiently to separate Ca^{++} components in an unambiguous manner. As part of her Doctoral Thesis research in my laboratory, Ellyn Wheeler-Clark examined the effects of La^{+++} on ^{45}Ca movements in rat brain area slices (22). It was found that La^{+++} inhibited uptake and subsequent binding of Ca^{++}, whereas decreased temperature markedly decreased ^{45}Ca extrusion (by inhibiting energy dependent ^{45}Ca efflux). Similar effects of La^{+++} and lowered temperature were also found in rat lung slices (23) and in murine pancreatic beta cells (24). Use of La^{+++} as a tool to identify differences in Ca^{++} metabolism was also valuable for us in identifying qualitative variations between aortic media-intimal strips and cultured aortic smooth muscle cells (25) and in demonstrating the lack of variation in ^{45}Ca uptake and retention and in contractile responses of rabbit aortic smooth muscle in tris or bicarbonate (phosphate-free) buffer solutions (26).

Numerous earlier studies of actions of La^{+++} conducted with different biological systems were cited as part of a later review article on lanthanides as probes for calcium in biological systems (27). One early study that was particularly relevant to us was Sanborn and Langer's characterization of the mechanism of action of lanthanum in mammalian cardiac tissue (28). They found that La^{+++} released ^{45}Ca from a contractile dependent Ca^{++} pool derived primarily from superficially located sites. This was in agreement with our earlier studies in smooth muscle (6-8), and with other studies I conducted at that time in frog sartorius (29) and rectus abdominis (30) muscles showing in all cases that the Ca^{++} displaced by La^{+++} was superficially located. A later interesting study with La^{+++} by Kramsch and associates (31) in rabbits fed atherogenic diets showed that treatment with La^{+++} did not lower the elevated serum cholesterol and Ca^{++} levels but did prevent the rise in arterial Ca^{++} content and suppress all symptoms of atherosclerotic lesions. Thus, in different types of muscle and in atherosclerosis, La^{+++} was able to act in a specific manner at a Ca^{++}-dependent step to alter the physiological (or pathological) tissue response.

Initial studies on isolated mitochondria demonstrated that La^{+++} inhibited mitochondrial Ca^{++} transport (32) by binding to the Ca^{++} sites and inhibiting the Ca^{++} carrier (33). We pursued studies of binding of La^{+++} to cellular membranes, primarily in an aortic smooth muscle microsomal preparation developed by Phyllis Kutsky. These studies with either La^{+++} (34) or Scatchard-coordinate plots (35) confirmed that La^{+++} binds on the membrane surface and decreases the number of sites available for Ca^{++}. Subsequently, we (36) found that the increase in La^{+++}-resistant low affinity Ca^{++} uptake elicited with stimulatory concentrations of high K^{+} in rabbit aortic smooth muscle was blocked by inhibitors of mitochondrial respiration (oligomycin, antimycin A, cyanide), indicating that (a) the K^{+}-induced uptake of Ca^{++} sequentially preceded the increased retention of La^{+++}-resistant low affinity Ca^{++} and (b) the dissociation of these two events by mitochondrial inhibitors occurs because the increased Ca^{++} is accumulated in the mitochondria.

10.7. Models Developed

A few years after the initial studies with La^{+++}, I wrote a review article on the cellular pharmacology of lanthanum (37), pointing out that use of techniques involving La^{+++} as a partial and specific antagonist of Ca^{++} would contribute greatly to resolution of specific Ca^{++}-dependent actions. Also, it was noted that employment of La^{+++}, particularly in smooth muscle, may well serve as a model for eventual development of additional agents acting at other sites in similarly defined Ca^{++}-dependent cellular actions. In the twenty years since that time, our knowledge of the roles of Ca^{++} in biological systems has become considerably more detailed and integrated. As my research on La^{+++}-related actions in smooth muscle progressed, I attempted to integrate my results (and those of others) in summary diagrams and cellular models. These figures were developed primarily to depict increasingly detailed concepts of drug-Ca^{++} interactions and the underlying methodology employed. However, they also show clearly how use of La^{+++} played an important role in defining these events.

The routes by which Ca^{++} appears to enter the intracellular compartment are schematically illustrated in Figure 3 (38). On the basis of specific inhibitory effects of different pharmacological agents, three Ca^{++} entry mechanisms as well as both superficially-located and intracellular high and low affinity Ca^{++} sites were visualized. The techniques we used to dissociate different Ca^{++} fractions are summarized in Table 2. Use of La^{+++} was clearly of primary importance for removal of extracellular bound Ca^{++} and for blockade of Ca^{++} fluxes. Similarly, the specificity of different Ca^{++} antagonists, including La^{+++}, for membrane and cellular channels and binding sites are summarized in Table 3. Dissociation of various loci with specific Ca^{++} antagonists provided the experimental basis for postulation of differing functional sites and channels. Finally, adding experimental information (39) and some of the inhibitory agents (40) to the schematic diagram in Figure 3 yielded the model for vascular smooth muscle shown in Figure 4 (to which La^{+++} has been added). Clearly, the different specific effects of the various inhibitors on entry and release of ^{45}Ca components help dissociate the postulated sites and channels.

Table 2. Dissociation of Various Ca^{++} Binding Sites and Uptake Mechanisms

Experimental Objective	Technique Employed
Removal of Extracellular Ca^{++}	Ca^{++} Deficient Solutions La^{+++}
Blockage of Ca^{++} Fluxes: Uptake Efflux	 La^{+++} Low Temperature + La^{+++}
Separation of Bound Ca^{++} Fractions	Scatchard Coordinate Plot Sr^{++}

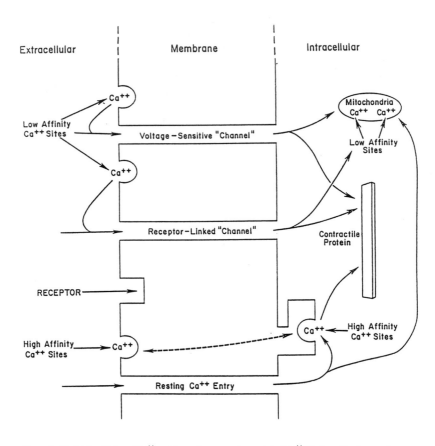

Figure 3: Model for different Ca^{++} uptake pathways and associated Ca^{++} binding sites in vascular smooth muscle. From (38)

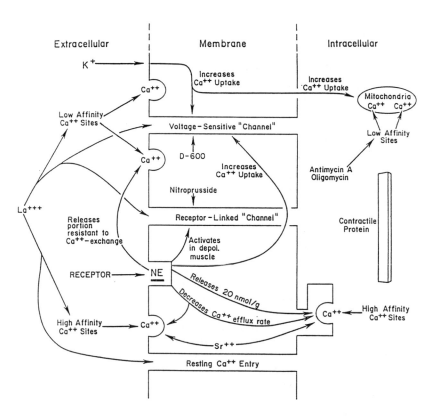

Figure 4: Model for sites of action of lanthanum and other inhibitory agents on Ca^{++}-uptake pathways and associated Ca^{++}-binding sites in vascular smooth muscle

Table 3 Sites/Channels at Which Different Types of Ca^{++} Antagonists Act

Sites/Channels	Blocking Agents
Superficial Ca^{++} Binding Sites (mainly low affinity)	Aminoglycoside Antibiotics La^{+++}
Resting Ca^{++} Entry	La^{+++}
Voltage-sensitive Ca^{++} Channel	Ca^{++}-blockers (e.g., Verapamil and Nifedipine, their derivatives) La^{+++}
Receptor-sensitive Ca^{++} Channel	Nitroprusside La^{+++}
Cellular Ca^{++} Binding Sites (microsome - high affinity)	Hydralazine Sr^{++}
Mitochondrial Ca^{++} Binding Sites (low affinity)	Antimycin A Oligomycin

10.8. Concluding Thoughts

The story developed was basically a kinetic one, deriving cellular models primarily from experimental data measuring the effects of stimulatory and inhibitory agents upon contractile responses and ionic fluxes. Use of La^{+++} provided the critical initial stimulus for these efforts because this ion was able to dissociate Ca^{++} components in a qualitatively and quantitatively unambiguous manner in smooth muscle in a way that was previously not feasible. Over time, a number of additional research techniques were developed and employed by many investigators to provide additional information about Ca^{++} distribution and mobilization patterns. These approaches included substances that combine with Ca^{++} (e.g., ion-specific chelators and intracellular dyes), sensitive techniques for measuring membrane currents (e.g., voltage and patch clamp in smooth muscle), and methods for observation of subcellular ion movements (e.g., specific ion probes). Extensive use of these newer techniques has provided considerable additional detailed information about the cellular physiology and pharmacology of smooth muscle. However, current concepts and knowledge in this very active research field are based upon and in general agreement with those obtained from earlier work in which use of La^{+++} was an important initial tool.

10.9. References

1.   Hurwitz, L., Battle, F. and Weiss, G.B.: Action of the calcium antagonists cocaine and ethanol on contraction and potassium efflux of smooth muscle. J. Gen. Physiol. 46:315-332, 1962.

2.   Weiss, G.B. and Hurwitz, L.: Physiological evidence for multiple calcium sites in smooth muscle. J. Gen. Physiol. 47:173-187, 1963.

3.   Shanes, A.M. and Bianchi, C.P.: The distribution and kinetics of release of radiocalcium in tendon and skeletal muscle. J. Gen. Physiol. 42:1123-1137, 1959.

4.   Lettvin, J.Y., Pickard, W.F., McCulloch, W.S. and Pitts, W.: Theory of passive ion flux through axon membrane. Nature (Lond) 202:1338-1339, 1964.

5.   Takata, M., Pichard, W.F., Lettvin, J.Y. and Moore, J.W.: Ionic conductance changes in lobster axon membrane when lanthanum is substituted for calcium. J. Gen. Physiol. 50:461-471, 1966.

6.   Weiss, G.B. and Goodman, F.R.: Effects of lanthanum on contraction, calcium distribution and Ca^{45} movements in intestinal smooth muscle. J. Pharmacol. Exp. Ther. 169:46-55, 1969.

7.   Goodman, F.R. and Weiss, G.B.: Dissociation by lanthanum of smooth muscle responses to potassium and acetylcholine. Am. J. Physiol. 220:759-766, 1971.

8.   Goodman, F.R. and Weiss, G.B.: Effects of lanthanum on ^{45}Ca movements and on contractions induced by norepinephrine, histamine and potassium in vascular smooth muscle. J. Pharmacol. Exp. Ther. 177:415-425, 1971.

9.   Van Breemen, C., Farinas, B.R., Gerba, P. and McNaughton, E.D.: Excitation-contraction coupling in rabbit aorta studied by the lanthanum method for measuring cellular calcium influx. Circ. Res. 30:44-54, 1972.

10.   Deth, R. and Van Breemen, C.: Relative contributions of Ca^{2+} influx and cellular Ca^{2+} release during drug induced activation of the rabbit aorta. Pflügers Arch. 348:13-22, 1974.

11.   Burton, J. and Godfraind, T.: Sodium-calcium sites in smooth muscle and their accessibility to lanthanum. J. Physiol. 241:287-298, 1974.

12.   Deth, R.C.: Effect of lanthanum and reduced temperature on ^{45}Ca efflux from rabbit aorta. Am. J. Physiol. 234:C139-C145, 1978.

13.   Brading, A. and Widdicombe, J.H.: The use of lanthanum to estimate the numbers of extracellular cation-exchanging sites in the guinea-pig's taenia coli, and its effects on transmembrane monovalent ion movements. J. Physiol. 266:255-273, 1977.

14.   Karaki, H. and Weiss, G.B.: Alterations in high and low affinity binding of ^{45}Ca in rabbit aortic smooth muscle by norepinephrine and potassium after exposure to lanthanum and low temperature. J. Pharmacol. Exp. Ther. 211:86-92, 1979.

15.   Karaki, H. and Weiss, G.B.: Effects of stimulatory agents on mobilization of high and low affinity site ^{45}Ca in rabbit aortic smooth muscle. J. Pharmacol. Exp. Ther. 213:450-455, 1980.

16.   Weiss, G.B. and Goodman, F.R.: Distribution of a lanthanide (^{147}Pm) in vascular smooth muscle. J. Pharmacol. Exp. Ther. 198:366-374, 1976.

17.   Weiss, G.B. and Goodman, F.R.: Interactions between several rare earth ions and calcium ion in vascular smooth muscle. J. Pharmacol. Exp. Ther. 195:557-564, 1975.

18.   Weiss, G.B.: Quantitative measurement of binding sites and washout components for calcium ion in vascular smooth muscle. In: Calcium in Drug Action, ed. by G.B. Weiss, pp.57-74, Plenum Press, New York, 1978.

19.   Weiss, G.B.: Calcium and contractility in vascular smooth muscle. Adv. Gen. Cell. Pharmacol. 2:71-154, 1977.

20.   Hester, R.K. and Weiss, G.B.: Comparison of degree of dependence of canine renal arteries and veins on high and low affinity calcium for responses to norepinephrine and potassium. J. Pharmacol. Exp. Ther. 216:239-246, 1981.

21.   Goodman, F.R., Weiss, G.B., Karaki, H. and Nakagawa, H.: Differential calcium movements induced by agonists in guinea pig tracheal muscle. Eur. J. Pharmacol. 133:111-117, 1987.

22.   Weiss, G.B. and Wheeler, E.S.: Inhibition of [45]Ca movements by lowered temperature or lanthanum in rat brain slices. Arch. int. Pharmacodyn. Thér. 233:4-20, 1978.

23.   Goodman, F.R. and Gardiner, T.H.: Characteristics of [45]Ca uptake and efflux in rat lung slices. Arch. int. Pharmacodyn. Thér. 230:31-41, 1977.

24.   Hellman, B.: Calcium and pancreatic β-cell function 3. Validity of the La^{3+}-wash technique for discriminating between superficial and intracellular [45]Ca. Biochim. Biophys. Acta 540:534-542, 1978.

25.   Kutsky, P. and Weiss, G.B.: Differing [45]Ca efflux responses in cultured canine aortic smooth muscle cells and canine aortic media-intimal strips. Arch. int. Pharmacodyn. Thér. 246:47-60, 1980.

26.   Karaki, H. and Weiss, G.B.: Rabbit aortic contractile responses and [45]Ca retention in tris and bicarbonate buffers. Arch. int. Pharmacodyn. Thér. 252:29-39, 1981.

27.   Martin, R.B. and Richardson, F.S.: Lanthanides as probes for calcium in biological systems. Quart. Rev. Biophys. 12:181-209, 1979.

28.   Sanborn, W.G. and Langer, G.A.: Specific uncoupling of excitation and contraction in mammalian cardiac tissue by lanthanum. J. Gen. Physiol. 56:191-217, 1970.

29.   Weiss, G.B.: On the site of action of lanthanum in frog sartorius muscle. J. Pharmacol. Exp. Ther. 174:517-526, 1970.

30.   Weiss, G.B.: Inhibition by lanthanum of some calcium-related action in frog rectus abdominis muscle. J. Pharmacol. Exp. Ther. 185:551-559, 1973.

31.   Kramsch, D.M., Aspen, A.J. and Apstein, C.S.: Suppression of experimental atherosclerosis by the Ca^{++}-antagonist lanthanum. J. Clin. Invest. 65:967-981, 1980.

32.   Mela, L.: Inhibition and activation of calcium transport in mitochondria. Effects of lanthanides and local anesthetic drugs. Biochemistry 8:2481-2486, 1969.

33.   Lehringer, A.L. and Carafoli, E.: The interaction of La^{3+} with mitochondria in relation to respiration-coupled Ca^{2+} transport. Arch. Biochem. Biophys. 143:506-515, 1971.

34.   Kutsky, P. and Goodman, F.R.: Calcium incorporation by canine aortic smooth muscle microsomes. Arch. int. Pharmacodyn. Thér. 231:4-20, 1978.

35.   Kutsky, P., Weiss, G.B. and Karaki, H.: Delineation of high and low affinity [45]Ca incorporation and of [45]Ca efflux in canine aortic smooth muscle microsomes. Gen. Pharmacol. 11:475-481, 1980.

36.   Karaki, H. and Weiss, G.B.: Inhibition of mitochondrial Ca^{2+} uptake dissociate potassium-induced tension responses from increased [45]Ca retention in rabbit aortic smooth muscle. Blood Vessels 18:28-35, 1981.

37.   Weiss, G.B.: Cellular pharmacology of lanthanum. Annu. Rev. Pharmacol. 14:343-354, 1974.

38.   Weiss, G.B.: Multiple Ca^{2+} sites and channels provide a basis for alterations in Ca^{2+} mobilization and vascular contractility. In: Vasodilation, ed. by P.M. Vanhoutte and I. Leusen, pp. 307-310, Raven Press, New York, 1981.

39.   Weiss, G.B.: Calcium kinetics in vascular smooth muscle. Chest 88S:200S-223S, 1985.

40.   Weiss, G.B.: Sites of action of calcium antagonists in vascular smooth muscle. In: New perspectives on calcium antagonists, ed. by G.B. Weiss, pp. 83-94, Williams-Wilkins, Baltimore, 1981.

CHAPTER 11

MEDICAL USES OF THE RARE EARTHS

C.H. EVANS
Department of Orthopaedic Surgery
University of Pittsburgh School of Medicine
Pittsburgh, PA   15261

11.1. Introduction

Metals have been used medicinally since antiquity; the ancient Chinese, for instance, ascribed therapeutic properties to gold. It is likely that Ekeberg, in 1797, was the first to suggest such an application for a rare earth. As Pyykkö and Orama describe in Chapter 1 of this volume, after repeating Gadolin's isolation of the new earth yttria Ekeberg noted that it had a sweet taste "almost like that of lead compounds but not so repulsive but rather contracting" and suggested that it could be "medically potent".

So far as we know, this idea lay dormant for over half a century until cerium oxalate burst upon the scene as an anti-emetic. Since then, there has been a more or less continuous flow of possibilities regarding the use of rare earths in medical practice. Few, however, have entered the clinic and fewer still became widely used.

It is neither necessary nor historically valuable to analyse every proposed medical application of a rare earth; readers interested in such matters can consult reviews by Ellis (1) and Evans (2). Instead this chapter concentrates on four areas where rare earths were applied to large numbers of individuals and thus influenced the manner in which patients were treated.

The first of these areas, which centres around the use of cerium oxalate as an anti-emetic, is of value for several reasons. Firstly, there was no obvious reason to use cerium oxalate for this indication; secondly, it gained unaccountably rapid and widespread endorsement as a treatment of not only sickness but also coughing and even nervous disorders; thirdly, its fall from favour was almost as rapid as it rise, and for reasons that are equally unclear. Understanding how all this came about could provide a window into the nature and practice of medicine in Victorian times.

C. H. Evans (ed.), Episodes from the History of the Rare Earth Elements, 205–228.
© 1996 *Kluwer Academic Publishers. Printed in the Netherlands.*

The second area to be examined in detail concerns the use of rare earths as antimicrobial agents. This application evolved from laboratory studies conducted in Germany at the end of the last century, which showed that various lanthanide salts prevented the growth of microorganisms. As described in this chapter, these observations were expanded by French investigators who, on the basis of weak *in vitro* data, began clinical trials for the treatment of tuberculosis. Surprizingly encouraging results were reported, but nothing further seems to have emerged from this line of treatment - why?

Rare earths have, however, returned to the clinic as topical antiseptic agents or use in burns. Yet the success of cerium nitrate in this regard may not lie so much with its antimicrobial properties, as with its ability to inactivate an immunosuppressive burn toxin. Here, perhaps, is a good case of the right drug being used for the wrong reason. Along similar lines, several complexes of the rare earths are used clinically in Hungary as anti-inflammatory agents. Yet the rationale behind their development, which postulates a strong link between inflammation and coagulation, is flawed. Nevertheless it provides an additional instance of where a drug originally developed for one clinical purpose finds better use for a different one.

The final example, that of the use of GdDTPA in magnetic resonance imaging, is dealt with rather briefly as it is a new application which remains under development; writing an historical account is probably premature. Nevertheless it is included as the only example where the rare earth compound was specifically designed for the clinical use to which it is being applied.

11.2. Cerium Oxalate in the Late 19th Century

11.2.1. INTRODUCTION

Sir J.Y. Simpson, M.D., F.R.S.E., Professor of Medicine and Midwifery in the University of Edinburgh, Scotland introduced the use of cerium salts as anti-emetic agents. In an address before the Medico-Chirurgical Society of Edinburgh reported the Monthly Journal of Medicine and Science of 1854 (3), good results were described for a variety of digestive disorders including "general chronic intestinal eruption", "irritable dyspepsia", "gastrodynia" (stomach pain), "pyrosis" (indigestion), "chronic vomiting" and the vomiting that accompanies pregnancy. Additional uses of cerium nitrate as a tonic and as an agent for use in convulsive diseases, such as chorea and epilepsy, were also mentioned. The full text of Simpson's 1854 paper is reproduced on the next page.

NOTE ON THE THERAPEUTIC ACTION OF THE SALTS OF CERIUM. BY J. Y. SIMPSON, M.D.—The salts of few metals were at present in our pharmacopœia, and Dr S. saw no reason why many more might not be added to their number. He had recently drawn the attention of the profession to the salts of nickel, which, as far as his observation went, presented much similiarity in action to those of iron and quinine, and seemed to be of use in cases of sick headache. He now proposed to read to the Society some very imperfect observations on the therapeutic action of some other metals: and first, as to cerium, which, given in the form of nitrate, and in one grain doses twice or thrice a-day, appeared to act as a sedative tonic of considerable value, strongly resembling bismuth, and the salts of silver. He had employed it in the first instance in cases of general chronic intestinal eruption—a peculiar and intractable form of disease, for which arsenic and nitrate of silver were generally prescribed, and where these remedies had failed, cerium had been tried with marked advantage. In irritable dyspepsia, with gastrodynia and pyrosis, and in chronic vomiting, its exhibition was attended with satisfactory results; and in the vomiting which occurs during pregnancy prompt relief was afforded. It was a good tonic, and a useful substitute for the salts of silver, bismuth, and hydrocyanic acid. Dr S. had not employed it much in convulsive diseases, as chorea and epilepsy, in which nitrate of silver was used, but the exhibition of the salts of cerium was certainly attended with this advantage, that it could be persevered in without any fear of discolouration of the skin. As far as his experiments with cadmium went, it bore much resemblance to the preparations of antimony, and excited diaphoresis and vomiting. Tellurium, besides its expense, was precluded from being used in practice by its disagreeable effects. Dr S. mentioned the particulars of a case where a dose had been inadvertently given to a student of divinity, and had been followed by the evolution of such a persistent odour, that for the remainder of the session the patient had to sit apart from his fellow-students.

Dr Gairdner (who occupied the chair during the reading of Dr Simpson's communications), in the name of the Society thanked Dr S. for his very interesting paper. He (Dr G.), however, felt doubtful how far a medical man was justified in experimenting with new remedies, before their effects had been ascertained on some of the lower animals. This query he threw out merely in the way of suggestion, in the hope of hearing the views of Dr Simpson and the other members of the Society on this interesting subject.

Dr Sellar said, that it was now a long time since the pharmacopœia had received any addition from the metals. Bismuth, he (Dr S.) believed, was the latest, and he thought that the profession was indebted to Dr Simpson for having opened up such a new and promising field of investigation. The question as to the propriety of instituting these experiments now ceased to exist, as the experiments had been successfully made, whether on inferior animals or not. He (Dr S.) had little faith in bismuth or nickel, and as to tellurium, the less it was given the better.

Dr Simpson thought, he had been justified in carrying out these experiments, as, with the exception of arsenic, none of the metals were poisonous in smaller doses than half a grain, and, in the present instance, he had previously tried the remedies in similar doses on his own person. He might mention that the cost of cerium was one penny a grain, and that in London, nitrate of silver was charged at the same price.

Simpson's 1854 paper as reported in the Monthly Journal of Medicine (3)

In a longer article, published in the Medical Times and Gazette of 1859, Simpson elaborated upon this theme, describing in some detail his experiences in treating the nausea and vomiting of pregnancy with cerium salts (4). By now he had switched from using the freely soluble cerium nitrate to the very insoluble cerium oxalate. This change was apparently due to the ready availability of cerium oxalate. As described in part I of this volume, the formation of insoluble oxalates was at that time a method commonly used in the purification of individual lanthanides. The preparations of cerium oxalate used in clinical practice would certainly have been contaminated with other lanthanides, especially the lighter members of the series such as lanthanum, neodymium and praseodymium, as well as some non-lanthanides like thorium. Simpson was of the opinion that any salt of cerium would be as pharmacologically effective as the oxalate.

Cerium nitrate, according to Simpson (4), was:

> "..of all individual remedies the simplest and surest agent that can be administered for the sympathetic vomiting of pregnancy ...I have been successful in curing vomiting in a larger proportion of cases than any other single remedy which I have used; and its good effects are not confined to the forms of vomiting which depend upon the sympathetic derangements of the stomach caused by changes, functional or pathological, in the uterus or other organs, but are manifested also in those forms of the disease which are due to different morbid conditions of the stomach itself..."

Despite this glowing testimonial, Simpson reminds the reader that cerium oxalate does not cure or alleviate vomiting in every case; however, greater success is to be expected with cerium oxalate than with any other drug existing at that time. Other anti-emetics in use during the mid-19th century included opium, bismuth salts, prussic acid, naphtha, creosote, carbonic acid gas (as present in champagne and other fizzy drinks), salicine, silver salts and lead acetate.

Simpson recommended a dose of cerium oxalate of 1-2 grains (1 grain = 65mg) three times a day, or more often, in the form of a pill or mixed with a few grains of gum tragacanth as a powder. The effect could be instantaneous, or might take several days.

11.2.2. WHY CERIUM?

Nothing that was known at that time about the biological properties of the lanthanides gave *a priori* justification to their use as anti-emetics or sedatives. Why Simpson should have employed cerium salts in this capacity is unknown.

The 1854 report on the therapeutic action of the salts of cerium begins with the observation that, "The salts of few metals were at present in our pharmacopoeia and Dr. S. saw no reason why many more might not be added to their number." Moreover a Dr.

Sellar, one of the discussants of Simpson's paper, agreed "that it was now a long time since the pharmacopoeia had received any addition from the metals. Bismuth, he believed, was the latest". Interestingly, bismuth salts were also used for the treatment of vomiting and, unlike cerium, continue to be used for this purpose today. There is thus the possibility that, at the time Simpson was beginning his work in this area, the use of metals as anti-emetic agents was in vogue. In addition to bismuth, salts of silver, iron, lead and nickel had already been used for similar purposes.

The contemporary use of metals to treat gastrointestinal disturbances may have engendered the experimental application of additional metals in this context, but does not explain the choice of cerium. Already half a century had passed since the discovery of this metal, (see Chapter 2) so the novelty factor can presumably be excluded. In reading the 1854 account (3), one is left with the feeling that Simpson may well have engaged in a fairly random trial of the salts of all metals that were cheap, readily available, lacking in obvious toxicity and absent from the pharmacopoeia.

11.2.3. PROFESSIONAL RESPONSES TO THE USE OF CERIUM

Members of the medical community were quick to repeat Simpson's experiences with cerium oxalate. According to Lee (5), within a few months of Simpson's initial report numerous trials of cerium oxalate's efficacy in different gastric affections were initiated both in Europe and America. Simpson's high standing in the medical community was one reason for such prompt and widespread attention. These early trials proved successful for both the vomiting of pregnancy and other gastric distrubances, prompting Lee comment that cerium oxalate "promises to assume a permanent place among the mineral tonics." Lee went on to describe eight cases where he had administered cerium oxalate for vomiting of various causes, including pregnancy, phthisis (tuberculosis), and "hysterical amenorrhoea", as well as one case in which vomiting was idiopathic. The almost uniform success in these cases led to trials on fourteen patients with atonic dyspepsia. Again, cerium oxalate proved highly effective (5). However, it failed to treat several cases of gastritis both idiopathic or secondary to "debauch" or *delirium tremens* (5).

The literature from this point on contains a number of complimentary articles from other physicians who had used cerium oxalate in a variety of difficult cases with dramatic success. Dr. Thomas McKie of South Carolina, for instance, recorded how in 1860 he was confronted by: "a most obstinate case of vomiting dependent upon uterine engorgement which had resisted all remedies usually adminstered in such cases (6)." As a result of Simpson's article, he administered cerium oxalate as a last resort. The patient received a grain of this substance three times a day for two days, increasing to 2 grains on the third day during which vomiting ceased. McKie's article goes on to describe how he had found cerium oxalate to be a useful anti-emetic both in children and other adult cases of vomiting, with a therapeutic effect sometimes occurring instantaneously and sometimes requiring up to several days.

Dr. H.W. Jones, writing in the Chicago Medical Journal of 1861 (7) reported success in four very difficult cases of vomiting during pregnancy, while a Dr. Busey is reported to have used cerium oxalate in conjunction with opium (8).

In 1878, Dr. Francis Edward Image recommended increasing the dose of cerium oxalate to 10 grains, commenting that "the official dose of 1-2 grains is as a rule so useless that the preparation has been stigmatized as the 'oxalate of mud'"(9). Of note is Image's comment that cerium oxalate "has of late years much fallen out of use" (see next section). Nevertheless, according to Image, he had "not met with a case in which nausea had not been very considerably relieved, and in most cases completely checked, by ten-grain doses of the oxalate of cerium."

Shortly thereafter, the successful use of cerium oxalate in sea-sickness was reported. According to Dr. W.H. Gardner, cerium oxalate administered at doses of up to 20 grains every 2 or 3 hours in about a tablespoonful of water "will cure more cases than champagne, bromide of potassium, choral, or anything else I have ever tried (10)." A success rate of 75% was claimed. In the same article Gardner also described dramatic results in a patient whose "continuous and exhaustive" vomiting resulted from a gunshot wound following "some little discussion over a game of cards." His symptoms disappeared in about 20 hours after taking 10 grains of cerium oxalate every 3 hours. "The man made a speedy and good recovery, and is still driving a stage-coach in Arizona."

In response to Gardner's article, a Dr. M.M. Waldron of Hampton VA wrote a letter published in the June 23rd, 1888 issue of the Medical Record, in which he confirmed the efficacy of cerium oxalate in sea-sickness(11). According to Waldron, 15 grains of cerium oxalate given every 2 hours, taken dry on the tongue, worked wonders on both coastal and transatlantic voyages.

Clearly, then, the medical literature of the times records a ringing endorsement of Simpson's claims. Although the evidence was what we would today condemn as anecdotal, by the standards of the times cerium oxalate had been independently put to the test on multiple, separate occasions and had proved itself. Scores of cases could be cited to support the efficacy of cerium oxalate in treating vomiting and other such digestive disturbances. By the end of the 19th century, cerium oxalate had found its way into the pharmacopoiae of a number of countries including the USA, Great Britain, Austria and Japan, with recommended doses ranging from 1-2 grains (65-130 milligrammes) in the British Pharmacopoia of 1882 to 1 gramme (just over 15 grains) in the Japanese Pharmacopoia. The Material Medica and Therapeutics for 1882 recommends 1-10 grains for adults and ¼-½ grain for infants and children under two. According to one pharmacy text book of 1874 (12), cerium oxalate was extensively prescribed in Europe and the United States.

A number of commercial, pharmaceutical preparations of cerium oxalate became readily and cheaply available. Because the chemistries of the individual lanthanides are so similar, these elements are difficult to separate one from another with formation of oxalates being frequently used at that time during the purification procedures. This made lanthanide oxalates common and cheap, but invariably contaminated with other elements, particularly other light lanthanides. One commercial preparation, known as Cerocol, was produced as tablets, each containing 50mg of colloidal cerium oxalate (13). The manufacturer, Coates and Cooper, was based in Middlesex, England but is no longer in business. Another pharmaceutical preparation of cerium oxalate, Novonaurin, was manufactured until recently (and may still be) in Spain (14).

11.2.4. CERIUM'S FALL FROM FAVOUR

Despite the foregoing effusion of enthusiasm, cerium oxalate fell into disuse as a pharmaceutical early in the 20th century. The reasons for this decline are not obvious. However, it would appear that not all practitioners obtained the same good results with this drug.

According to Baehr and Wessler (15), "diversity of opinion regarding the therapeutic value of cerium oxalate had existed ever since the introduction of the drug in 1854 by Simpson before the Medico-Chirurgical Society of Edinburgh." Despite the fact that many "subsequent observers recorded their endorsement of the use of cerium oxalate for the relieve of all sorts of gatrointestinal disturbances...many others have been absolutely unable to achieve any alleviation of the identical disorders by the exhibition of this substance." Evaluation of these statements is difficult, as Baehr and Wessler give no references to indicate the source of these negative observations. Neither have I come across any published data refuting an anti-emetic effect in humans. Perhaps then, as now, it was difficult to publish negative data. However, as noted in the previous section, the 1878 paper of Image (9) did contain the comment that cerium oxalate had "of late years" fallen out of favour. No further explanation was given.

According to Baehr and Wessler (15), the New York County Medical Society convened a special committee to look into this matter. However, the reference they cite (16) is to an evaluation of the use of cerium oxalate to relieve cough, and I have been unable to identify any report from the committee on the subject to which these authors refer. Nevertheless, Baehr and Wessler infer that the committee concluded that the different cerium oxalate preparations on the market were not of the same composition and, therefore, not of equal value.

The main purpose of Baehr and Wessler's article was to report studies they had performed on dogs. The first set of experiments were toxicological and involved feeding oxalates of various lanthanides, including cerium, to the animals. They found that in dogs, as in humans, these substances were harmless. The investigators then administered commercial preparations of cerium oxalate to the animals which were induced to vomit by application of apomorphin hydrochloride, an emetic inducing reflex vomiting as occurs in pregnancy, or ipecac, an emetic acting by local irritation of the gastric mucosa. The results indicated that cerium oxalate had no influence on reflex vomiting, but that it did inhibit vomiting due to local gastric irritation. Baehr and Wessler point out that these observations are consistent with the inability of the body to absorb cerium oxalate from the gastrointestinal tract. Cerium oxalate probably inhibits local gastric irritation by forming an insoluble protective coating to the stomach and thus acts in a manner analogous to bismuth subnitrate.

Baehr and Wessler also tested the effects of cerium nitrate, the salt reported in Simpson's original paper of 1854. The results of these experiments were even less encouraging, with orally administered cerium nitrate causing nausea, vomiting, violent retching, diarrhoea and tenesmus.

The degree to which these data, published in the academic literature, led to the eventual disuse of cerium oxalate as an anti-emetic is difficult to establish. No subsequent publications on this topic have appeared in the scientific literature, and the clinical literature maintains a similar silence. Be this as it may, cerium oxalate fell out of favour as an anti-emetic early in the 20th century and remains unprescribed today.

11.2.5. CERIUM OXALATE FOR THE RELIEF OF COUGHING

By the 1880s, reports were beginning to appear suggesting that cerium oxalate was not only a useful treatment for vomiting, but also for coughing. A Mr. Thomas Clark is cited by Dr. Image writing in The Druggist Circular and Chemical Gazette of 1878 (17) as noting that:

> "A lady has suffered from some years with cough and difficulty
> of breathing on the least exertion, the outcome of an acute
> attack of pneumonia, the cough being most troublesome in the
> morning on getting up; so bad as to cause sickness."

After taking 5 grains of cerium oxalate each day, half-an-hour before rising, the patient showed "less noise in breathing, less abdominal action, no cough in the morning, and increased strength." Success was also reported in treating consolidation of the right lung and was alluded to in a variety of additional cases.

Dr. Gardner, in his article on the use of cerium oxalate in seasickness (10), also mentions its ability to relieve the cough of phthisis (tuberculosis); a similar conclusion had been reached earlier by Cheesman (18), although not all patients responded. In Cheesman's trial of 69 patients with phthisis, cerium oxalate was markedly effective in 39 patients, moderately so in 19 patients and ineffective in the remaining 11 patients.

A committee of the New York County Medical Society met to consider the use of cerium oxalate in treating coughing and published the following conclusions (16):

1. Cerium oxalate may be given safely in doses of ten grains or more three times a day, for many days in succession.

2. The only symptom noted from such doses is a slight dryness of the mouth for the first few days.

3. It is probably more efficient when taken dry upon the tongue.

4. Its effects are not fully apparent until it has been taken two or three days and continue about the same length of time after its use is suspended.

5. For chronic cough it is best taken on an empty stomach early in the morning and at bed-time, with other doses during the day, if required, the initial dose for an adult being five grains.

6. It is in the majority of cases an efficient cough medicine, at least for a considerable time, and it is very valuable as an alternate with other drugs used for that purpose.

7. It does not disturb the stomach as do opiates and most other cough remedies; but on the contrary it tends to relieve nausea and to improve digestion.

8. The different preparations on the market are not of equal value, and when success is not obtained with one, another should be substituted.

The eighth conclusion suggests that not all physicians had found cerium oxalate of use in treating coughs. However, although the effects of cerium oxalate on coughing were not uniformly reported as strongly encouraging, it was generally agreed to be superior to other cough medicines in not deranging the digestion.

The supposed ability of cerium oxalate to prevent coughing was thought to reside with its actions as a sedative. In his 1854 paper, Simpson reported the use of cerium nitrate in convulsive diseases such as chorea. Indeed, Simpson is reported as considering cerium oxalate "as a good nerve tonic and almost a specific in chorea" (19). Dr. Ramskill (20) recorded two cases of epilepsy preceded by a "gastric aura" which were benefited by cerium oxalate, whereas cases lacking this aura were not. (Gastric aura was described as a "sense of faintness, and of something turning upside down at the epigastrium.")

Unlike its use in the treatment of vomiting, the application of cerium oxalate in coughing and convulsive diseases never gained wide acceptance and did not enter pharmacopoiae in this context.

11.3. Rare Earths As Anti-Microbial Agents

11.3.1. INTRODUCTION

Work conducted in Germany at the end of the 19th century demonstrated that lanthanides inhibit the growth of microorganisms (21,22). In his paper describing the antibacterial effects of cerium, didymium, lanthanum and yttrium nitrates, Drossbach

(22) infers that the antiseptic properties of lanthanides were already known, but that their widespread use in this regard was constrained by their high price. (This contrasts with Simpson's earlier reference to their cheapness - perhaps the market price had risen steeply or simply differed between Great Britain and Germany). However, Drossbach went on to note that the price of these chemicals had fallen due to their production as a by-product of glassmaking. Because of the close chemical similarities between various different lanthanides, their chemical separation would not be necessary; furthermore, lanthanides were not known to be toxic to mammals. Thus, according to Drossbach, they should be tested clinically in diseases resulting from microbial infections. It is noteworthy that this suggestion came shortly after the germ theory of disease was becoming widely accepted, and Koch's postulates were becoming the diagnostic criteria for microbial diseases. Koch himself had suggested that the salts of certain metals, especially gold, had anti-bacterial properties and could thus be of clinical utility.

11.3.2. EARLY PRE-CLINICAL EXPERIENCES - CONTRIBUTIONS BY FROUIN

During the early decades of the 20th century, further laboratory studies conducted in Germany, France and Italy confirmed and extended knowledge of the antimicrobial properties of the rare earths. In general, the heavier lanthanides were found to be stronger inhibitory agents than the lighter lanthanides, and bacteria were more sensitive than fungi.

Albert Frouin, working in France, published a series of communications on his experiments in which lanthanides were added to various bacilli (23-28). These experiments reflected Frouin's interest in the mineral requirements of bacteria and in particular, that of magnesium. In his first experiments (23), Frouin attempted to replace the magnesium requirement of *Bacillus pycocyanique*, a bacterium producing a characteristic pigment, with rare earths. During these studies the sulphates of neodymium, praseodymium and samarium were found to inhibit cell growth.

In the following paper (24), these observations were extended to the tubercle bacillus. Here the results were less uniform. Sulphates of cerium, lanthanum, neodymium, praseodymium and samarium stimulated growth of the cultures when administered at low doses (50mg/l). Neodymium and praseodymium sulphates at the high dose of 1g/l completely inhibited growth.

Photograph reproduced from Frouin, 1912 (24) showing the effect of praseodymium sulphate on the growth of the tubercle bacillus.

In 1914, Frouin and Roudsky reported their studies on the effects of lanthanide injections in guinea pigs which had been given cholera (26). All of 10 animals injected with double the minimum lethal innoculum of the microorganism survived when injected with lanthanum sulphate. When the innoculum was increased to 3½ times the lethal dose, 8 out of 10 died, although their survival was increased from 8 to 24-36 hours. In further experiments two apes of the genus *Cynomolgus* were also given a lethal innoculum of cholera, and survived following subsequent administration of lanthanum sulphate after a delay of 6 hours. Delaying treatment until 12 hours after inoculation did not prevent death in a further two animals. Thorium sulphate, which was also tested in these experiments, appeared to have a stronger effect against cholera.

In later experiments, sulphates of erbium, yttrium, lanthanum and thorium were administered intravenously to rabbits with dysentery (27). Although the erbium and yttrium salts showed the strongest anti-bacterial actions *in vitro*, they were unable to prevent the death of the infected rabbits into which they were injected. Injections of lanthanum and thorium sulphates, in contrast, allowed the animals to survive.

Frouin returned to the tubercle bacillus in a 1923 (28) paper in which he examined the effect of pH on the inhibitory properties of the rare earths. He found that when the pH values of media were adjusted to pH 6.2-7, rare earths stimulated bacterial growth; in unadjusted media of pH 5.2-5.6, lanthanides were inhibitory. It is possible that, unknown to Frouin, the rare earths precipitated from solution at the higher pH values (29).

On such flimsy evidence clinical trials in patients with tuberculosis were initiated. The use of the rare earths in this manner may have been encouraged by the earlier use of gold salts to treat this disease.

11.3.3. USE OF RARE EARTHS IN TUBERCULOSIS

Based upon Frouin's *in vitro* evidence that lanthanide sulphates could, at the appropriate pH and if given at high enough concentrations, inhibit the growth of the tubercle bacillus, human clinical trials were initiated by the French workers Drs. H. Grenet, H. Drouin, M. Brou and M. Esnault (30,31).

In a paper presented to the "Société Médicale des Hopitaux" in May, 1920, Esnault and Brou (30) described their results obtained the previous year by administering rare earth salts to patients with tuberculosis. Patients were given a series of 15-20 daily, intravenous injections of a 2% solution of neodymium or samarium sulphate in an escalating dose of 1-5 mls per injection. Each series of injections was separated from the next by a 15 or 20 day rest. During their treatment, patients continued with the "diététo-hygénique" cure, but took no other medication.

Interpretation of the data is considerably confounded by the fact that patients were also given 2 injections per week of bismuth, a more common treatment for tuberculosis at that time. Esnault and Brou commented that they now used bismuth only rarely because of its toxicity and because the rare earths gave, in general, equally good results.

In summarizing their findings, the authors noted that injections of lanthanide sulphates produced no adverse effects other than an occasional increase in temperature of 0.3-0.4°. Bismuth, on the contrary, often gave an immediate pain in the jaws and a taste of blood in the mouth. Of the 20 patients they treated, 5 patients got worse, 12 improved markedly and 3 improved slightly. By the third series of injections, occurring at the end of 2 months of treatment, there was a general improvement in the health of the patients, with a return of appetite, increase in weight, reduced purulence and quantity of spit. Furthermore, the bacteria themselves showed altered morphologies and staining behaviour.

The authors pointed out that the number of patients showing improvement was impressive, given the severity of the disease. They were continuing with the lanthanide treatments in the firm hope of effecting a cure. Furthermore they expected the curing of patients to become the rule once treatment was initiated early in the disease process, rather than in the advanced cases used for this trial.

This optimism was echoed by Grenet and Drouin in a paper read before the 14th French Congress of Medicine held in Brussels in May, 1920 (31). Fifty-two cases were reported, including the 20 of Esnault and Brou as well as 6 additional patients of theirs which had only received a single series of injections. This compilation is a little suspect, as the patients of Grenet and Drouin were injected with sulphates of samarium, neodymium and praseodymium only, whereas those of Esnault and Brou had, as noted above, also received bismuth. Be this as it may, all patients were reported to have tolerated the injections well with, perhaps, a small, temporary increase in the temperature of some patients.

Grenet and Drouin reported that, as a result of these injections, the overall state of the patients improved, including increased appetite and weight gain. Coughing and spitting were reduced, and radiological auscultation improved. Bacterial examination of sputum revealed marked morphological changes. Furthermore, when these bacteria were injected into guinea pigs, they appeared less virulent.

Although the authors did not claim to have cured all patients, they mentioned that injections were given on an outpatient basis such that patients could continue working and certain individuals who had to give up working were able to start again; one such individual was a glass-blower and another a baker who worked nights.

In explaining the mode of action of the rare earths, Grenet and Drouin not only made reference to the original *in vitro* work of Frouin showing a direct antimicrobial effect, but also mentioned more recent, unpublished, work by the same investigator showing that injection of lanthanides into experimental animals induced a mononucleosis. The increased abundance of mononuclear cells would presumably aid the biological destruction of the microorganisms. Lastly, they suggested that the rare earths promoted sclerosis and hence tissue repair.

On this high note, the literature on this subject abruptly ends. According to Arvela (32), lanthanides have also been used to treat leprosy, but no further information on this subject is available. We are left to wonder whether further trials failed to confirm these promising early results, or whether the advent of newer, more potent anti-bacterial drugs, such as the sulphonamides, overtook the field.

11.3.4. INTROCID

Following the work of a variety of German investigators, a preparation of cerium iodide known as "Introcid" was introduced in the 1920s as an antimicrobial agent of particular help in treating puerperal fever. Peters (33) reported trials in which Introcid showed prophylactic activity against sepsis in all of 100 cases tested, although it had no therapeutic activity in 6 patients with established sepsis, all of whom died. The preparation was reported as non-toxic, apart from slight thrombosis at the injection site. As an additional advantage, Introcid was shown to be sterile and stable upon storage. Peters suggested that the antiseptic effects of introcid were potentiated by the radioactivity of cerium. (Cerium was widely thought to be radioactive at this time, a mistake arising from its common contamination with the radioactive element thorium).

Introcid is no longer used and, like many other such preparations, has disappeared from the literature.

11.3.5. CERIUM IN BURNS

Infection is a frequent complication of burns, leading to considerable morbidity or even death. As antibiotics do not reduce the incidence of septic death in burn victims, other antiseptic agents have been tested.

Silver nitrate, which had been used since antiquity as an antiseptic agent, was introduced by Moyer in 1945 for the local antimicrobial treatment of burns. Clinical trials confirmed its effectiveness in this respect, but also revealed a major disadvantage. Because silver ions are so easily precipitated from solution, the silver nitrate had to be dissolved in distilled water. The resulting solution is highly hypotonic thus causing osmotic stress with loss of electrolytes. Silver nitrate had the additional problem of staining everything with which it came into contact, including skin and bedclothes, black. To circumvent these problems, Fox and his colleagues synthesized silver sulphadiazine as an improved agent for the topic treatment of burns (34). Such was its success that it rapidly became the treatment of choice.

Silver sulphadiazine, however, was not a panacea. It was only weakly active against certain gram-negative bacteria, and did not consistantly suppress bacterial growth in large burns covering more than 50-60% of the body's surface area. There was thus the need for additional antimicrobial activity; it was Monafo who suggested the use of cerium. He writes

> "in view of the sucessful introduction of silver nitrate for burn treatment, it seemed worthwhile to try to further increase the efficacy of topic silver solutions if possible by using another, hopefully equally non-toxic, metal that might have antibacterial properties. The lanthanides soon came under consideration after looking at the literature...Cerium was selected after some *in vitro* work and its relative absence of toxicity (35)."

The first clinical trials of cerium nitrate as a topical antiseptic for extensive burns reported dramatic results, especially when the cerium salt was combined with silver sulphadiazine - the use of cerium nitrate was accompanied by a 50% reduction in the anticipated death rate (36).

Subsequent trials at other centres led to a variety of outcomes ranging from endorsement to rejection of cerium nitrate's efficacy (e.g. 37-41). One study even suggested that cerium interfered with the antimicrobial properties of silver sulphadiazine (42). Nevertheless, the cerium nitrate-silver sulphadiazine combination gained a somewhat limited aceptance. A cream combining these agents is marketed by the Duphar Company in several European countries including Belgium, Holland, France and Germany, although it is not yet approved by the FDA in the USA.

Although it was the automicrobial activities of cerium nirate that led to its topical use in burns, the manner in which it improves the survival of burn victims may be more complex. Patients with severe burns suffer systemic disturbances that cannot easily be explained by damage to the skin alone. For over a century it has been suggested that the burnt skin releases substances which suppress normal physiological functions. Recent evidence confirms that burnt skin releases a "burn toxin" which, among other things, has an immunosuppressive effect. Cerium appears to interfere with the production or release of this toxin, and thereby to relieve immunosuppression (43,44). Burn patients entering the Kantonspital in Basel, Switzerland are now routinely bathed for 30 minutes in 0.04M cerium nitrate at the time of admission. This procedure provides an exceptionally high incidence of patient survival (45,46). A recent issue of the journal *Burns* has been entirely devoted to the topic of cerium nitrate in treating burns (47)

By a completely independent route, scientists in Hungary have produced "phlogosam",an ointment containing 3% sodium disulphosalicylate samarium anhydride (48). Manufactured by the Chemical Works of Gedeon Richter in Budapest, this ointment is used to treat, among other things, superficial burns. The logic behind this is quite different from that leading to the development of cerium nitrate creams in the USA. Phlogosam is a broad spectrum, topical anti-inflammatory agent whose usefulness in burns is thought to be related to its anti-inflammatory properties. This compound is described in greater detail in the next section.

11.4. Lanthanides as Anti-coagulant and Anti-inflammatory Agents

11.4.1. ANTI-COAGULANTS

The anticoagulant properties of the lanthanides were discovered in the 1920s by researchers at the Institute of Pharmacology at the University of Florence in Italy, as part of a larger research programme to investigate the pharmacology of the rare earths (49,50). These observations were seized upon by various German workers, particularly Erich Vincke and his collaborators at the University of Hamburg who embarked upon a systematic study of different salts of the various rare earth elements, concluding that their anti-coagulant properties resided with an inhibitory effect on prothrombin (51-54). Clinical application of these findings was complicated by the contemporary introduction of heparin. To compete with heparin, anticoagulants based upon the rare earth elements would need to be cheaper and simpler to administer. Moreover, intravenous injection of rare earth salts into human volunteers revealed that their benefits as anticoagulants were outweighed by their toxic side effects. These included chills, fevers, muscle pains, abdominal cramps, haemoglobinemia and haemoglobinuria (55).

Vincke and his collaborators searched for less toxic anticoagulant complexes of rare earths, finally introducing neodymium 3-sulpho-isonicotinate for human use in 1950 (56). Given the name "Thrombodym", this drug was administered intravenously at an initial dose of 250-375mg followed by a 125mg dose given every 12 hours. According to human studies, reported primarily in the German literature, the drug was well tolerated and both prevented and cured thromboses (57-60). Although its action was less immediate and less dramatic than that of heparin, the action of Thrombodym was longer lasting.

With such favourable properties, Thrombodym was administered to large numbers of patients in the 1950s, particularly in Germany where it had been developed, and was produced by C.F. Asche & Co. of Hamburg-Altona. Other rare earth complexes, such as the didymium salt of β-acetylproprionic acid (Helodym 88) and the cerium salt of p-aminobenzenesulphonic acid were also developed for human, anti-thrombotic use (61).

Despite this accumulated head of steam, the human use of these compounds diminished, and they are no longer used as clinical anticoagulants. The reasons for this decline are unrecorded, but they are likely to reflect the low cost, ease of application and wide acceptance of heparin. In cases where use of heparin is inappropriate, coumaric agents are available as alternative, orally active anti-thrombotic agents.
Although lanthanides do not find wide clinical application as anti-coagulants, they continue to generate interest as research tools with which to investigate the biochemistry of blood clotting (e.g. 60,61).

11.4.2. ANTI-INFLAMMATORIES

Through a strange twist of events, the anti-coagulant activities of the rare earths led to their use as anti-inflammatory agents. Topical anti-inflammatory preparations are presently produced and used in Hungary for a variety of indications.

Professor Jancso, working in Hungary at the Pharmacological Institute of the Medical University of Szeged, developed the hypothesis that the mechanism of inflammation involved blood clotting. According to Jancso: "we are convinced that some kind of a clotting process is involved in the mechanism of inflammation, and plays a central role in the causation of the inflammatory symptoms (62)."

To test his hypothesis Jancso tested various anticoagulants in his rat model of inflammation. To his surprise, heparin had no effect, but rare earth salts, Helodym 88 and Thrombodym were effective. However, both Helodym and Thrombodym were toxic and unless injected, extremely slowly, produced respiratory arrest, convulsions and death. (These side effects were not noted by investigators developing these agents as anti-coagulant agents for human use).

For this reason, Jancso sought new compounds of the rare earths which had anticoagulant properties yet were free from acute toxicity. He identified rare earth complexes of pyrocatechol sodium disulphonate as suitable agents. These proved to be potent anti-inflammatory agents in rats challenged by injections of a variety of pyrogens into the foot pad (64).

Based upon Jancso's findings and theories, Szporny examined the antiinflammatory properties of a number of different rare earth complexes (65). He found a 2:1 ratio of sulphosalicylic acid: Sm^{3+} to possess the most favourable pharmacologic properties, with an activity comparable to prednisolone ointment. Used as a solution, this preparation is marketed as "Phlogosol" with the composition 0.9g sodium disulphosalicylatosamarium anhydride and 0.03g hexachlorophenum in 30ml propylene glycol solution (66). As an ointment it is marketed as "Phlogosam"(48). Both preparations are produced by the Chemical Works of Gideon Richter in Budapest.

According to reports in the Hungarian literature, these preparations have wide applicability to a variety of inflammatory disorders when administered topically or by gargling. Present indications are shown in Table 1. Professor Jancso passed away several years ago and further research in this field has stopped.

Table 1 - Anti-inflammatory Preparations Containing Rare Earths (67)

FORMULATION	INDICATIONS
Phlogosam unguent	Various forms of dermatitis, Grade I and II burns, Adjuvant therapy of superficial thrombophlebitis
Plogosam foam	As above: a mild antiinflammatory and healing potion for skin
Phlogosol solution	Inflammations of the mouth, gums, tongue, larynx, etc

11.5. Nuclear Magnetic Resonance Imaging

Since the early 1970s there has been interest in using lanthanides as contrast agents to aid various imaging techniques used in clinical medicine. Most have been based upon the use of radioactive lanthanides for scintigraphy (68) or paramagnetic lanthanides for contrast enhancement in magnetic resonance imaging (MRI). An alternative strategy has been to use lanthanides as electron-dense materials to enhance the images formed by computed tomography (69). It is only as contrast enhancing agents in MRI that lanthanides have entered routine clinical use.

The principles underlying the use of contrast enhancing agents in MRI are quite straightforward. Images produced by MRI reflect local differences in T_1 and T_2 relaxation times. Substances which increase these differences provide a sharper image and improve sensitivity. Various radicals and paramagnetic ions have received attention in this context. The first application of the latter agents was published in 1982 (71) and involved the use of $MnCl_2$ to enhance *in vitro* MRI images of infarcted dog hearts. Of all the paramagnetic metals, however, gadolinium has the strongest T_1 relaxation properties and was thus the obvious preferred choice of contrast agent.

As early as 1982, Caillé et al (71) had injected solutions of $GdCl_3$ into rats and measured alterations in the T_1 times of several organs. However, simple salts of gadolinium are not well suited to human use, as Gd^{3+} ions are not cleared efficiently from the body; a chelate of Gd^{3+} which would promote rapid and complete elimination from the body was needed. The complex formed between Gd^{3+} and diethylenetriamine-N,N,N',N',N''-pentaacetic acid (DTPA) was found to fulfill this requirement.

Two groups simultaneously, but independently, developed GdDTPA as an MRI contrast medium. One of these groups was headed by Hanns-Joachim Weinmann at Schering A.G. in Berlin, and the other by Robert Brasch at the University of California, San Francisco. The German group injected the first human volunteer with GdDTPA in November, 1983 and the first literature reports on the use of GdDTPA in humans and experimental animals appeared in 1984 (72-75).

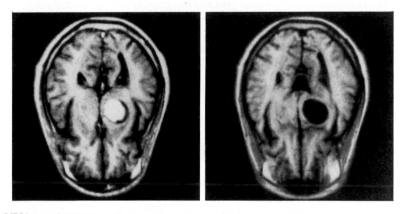

MRI image of cerebral tumour before (left) and (right) injection of Gd-DTPA (From ref. 75, with permission)

Groups of investigators led by Val Runge at Vanderbilt University and Gerald Wolfe at the University of Pennsylvania were also active in this area during its early development (76-79).

Patents for GdDTPA were filed by Schering, A.G. and by Brasch and Englestad. Both patents issued, but the United States patent issued to the second group was withdrawn in view of the slightly earlier filing date of Schering. This patent (80) covers the use of not only GdDTPA but also all paramagnetic metal chelates as MRI contrast media.

Acceptance of paramagnetic contrast agents was delayed by two factors. Firstly, most MRI experts were of the opinion that the intrinsic resolving power of MRI was such that contrast agents would never be required. Secondly, the use of such substances was viewed by some as depriving MRI of one of its main advantages, namely non-invasiveness. Nevertheless, the striking improvements in the quality of the images that were produced spoke for themselves, and GdDTPA has become extremely widely used.

11.6. Acknowledgements

I am indebted to Dr. William Monafo for his advice on the background to the use of cerium nitrate to treat burn patients and for permission to quote from one of his letters to me. Dr. George Lazar kindly provided me with information on the use of lanthanide-based anti-inflammatory agents in Hungary. Additional information on the development of GdDTPA for use in MRI was supplied by Drs. Gerald Wolf and Bob Runge and by Schering A.G. The efforts of Jerry Knab of Photo 24 in obtaining the photograph of Simpson's 1854 paper are very much appreciated. I am grateful to Mrs. Lou Duerring for her expert typing of this manuscript.

I would be very pleased to hear from readers who can provide additional insight into matters raised in this chapter.

11.7. References

1.   Ellis, K.J.: The lanthanide elements in biochemistry, biology and medicine. Inorganic Perspectives in Biology and Medicine 1:101-135, 1977.

2.   Evans, C.H.: Past, present, and possible future clinical applications of the lanthanides. Chapter 9, in Biochemistry of the Lanthanides, (Evans, C.H.) pp. 391-425, Plenum Press, New York, 1990.

3.   Simpson, J.Y: Note on the therapeutic action of the salts of cerium. Monthly Journal of Medicine p. 564, 1854.

4.   Simpson, J.Y. Clinical lectures on the diseases of women. Lecture XXII On spurious pregnancy - its prognosis, pathology and treatment. Medical Times and Gazette 40:277-281, 1859.

5.   Lee, C. On the therapeutical use of the oxalate of cerium. American Journal of the Medical Sciences 40:391-394, 1860.

6.   McKie, T.J. Oxalate of cerium. New Orleans Medical and Surgical Journal pp 746-747, May 1867.

7.   Jones, H.W. Observations on the therapeutic action of the oxalate of cerium in the vomiting of pregnancy. Chicago Medical Journal 4:65-69, 1861.

8.   Anon: Oxalate of cerium and caffein as preventatives of the nauseating effects of opium. Extracts from British and Foreign Journal pp. 214-215, 1878.

9.   Image, F.E.: On the employment of the oxalate of cerium in pregnant sickness. The Practitioner pp. 401-402, June, 1878.

10.   Gardner, W.H.: Oxalate of cerium in sea-sickness and other disorders. The Medical Record
        33:608, 1888.

11.   Waldron, M.M.: Oxalate of cerium in sea-sickness. The Medical Record 33:704, 1888.

12.   Parrish, E.: A treatise on pharmacy (4th Ed.) pp. 223-224, Henry C. Lea, Philadelphia, 1874.

13.   Martindale, The Extra Pharmacopoeia (R.G. Todd, Ed.) p. 1514, The Pharmaceutical Press,
        London, 1967.

14.   Martindale, The Extra Pharmacopoeia (J.E.F. Reynolds, Ed.) p. 1555, The Pharmaceutical Press,
        London, 1989.

15.   Baehr, G. and Wessler, H.: The use of cerium oxalate for the relief of vomiting: an experimental
        study of the effects of some salts of cerium, lanthanum, praseodymium, neodymium and thorium.
        Archives of Internal Medicine 2:517-531, 1909.

16.   Anon.: Oxalate of cerium to relieve cough. The Druggists Circular and Chemical Gazette
        24:166, 1880.

17.   Image, F.E.: The oxalate of cerium in pregnant sickness. The Druggist and Chemical Gazette
        22:170, 1878.

18.   Cheesman, H.: On the use of cerium oxalate for the relief of cough. The Medical Record p. 664,
        1880.

19.   Stille, A., Maisch, J.M., Caspari, C. and Maisch, H.C.C.: The National Dispensatory p. 1491,
        Lea Brothers & Co., Philadelphia, 1894.

20.   Ramskill, cited in: C.D.F. Phillips, Materia Medica and Therapeutics. Inorganic Substances.
        Volume II p. 115, William Wood & Co., New York, 1882.

21.   Bokorny, T.: Toxicologische Notizen über einige Verbindungen des Teller, Wolfram, Cer,
        Thorium. Chemiker-Zeitung 18:1739, 1894.

22.   Drossbach, G.P.: Über den Einfluss der Elemente der cer- und zircongruppe auf das Wachstum
        von Bakterien. Zentralbl Bakteriol. Parasitenk Abt. I Orig 21:57-58, 1897.

23.   Frouin, A. and Ledebt, S.: Action du vanadate de soude et des terres rares sur le développement
        du bacille pyocyanique et la production de ses pigments. Comptes Rendus de l'Academie des
        Sciences 72:981-983, 1912.

24.   Frouin, A.: Action des sels de vanadium et de terres rares sur le développement du bacille
        tuberculeux. Comptes Rendus de l'Academie des Sciences 72:1034-1037, 1912.

25.   Frouin, A.: Action du sulfate de lanthane sur le développement du B. Subtilis. Comptes Rendus
        de la Société de Biologie 74:196-197, 1913.

26.   Frouin, A. and Roudsky, D.: Action bactéricide et antitoxique des sels de lanthane et de thorium
        sur le vibrion cholérique. Action thérapeutique de ces sels dans le choléra expérimental.
        Comptes Rendus de l'Academie des Sciences 159:410-413, 1914.

27.   Frouin, A and Moussali, A.: Action des sels de terres rares sur les bacilles dysenteriques.
        Comptes Rendus de la Société de Biologie 82:973-975, 1919.

28.   Frouin, A and Guillaumie, M.: Nutrition minerale du bacille tuberculeux. Action favorisante ou empechante des sels de terres rares et des sels de fer. Comptes Rendus de la Société de Biologie 89:382-383, 1923.

29.   Evans, C.H.: Chemical properties of biochemical relevance. Chapter 2, in Biochemistry of the Lanthanides (Evans, C.H.) pp. 9-46, Plenum Press, New York, 1990.

30.   Esnault, M. and Brou, M.: Résultats du traitement de quelques cas de tuberculose pulmonaire chronique par les sulfates de terres rares. Société Medicale des Hospitaux pp. 606-615, May 7, 1920.

31.   Grenet, H. and Drouin, H.: Les sels de terres rares de la série du cérium. Traitement de la tuberculose pulmonaire chronique. Gazette des Hopitaux pp. 789-791, 1st and 3rd June, 1920.

32.   Arvela, P.: Toxicity of rare earths. Progress in Pharmacology 2:71-114, 1979.

33.   Peters, R.: Erfahrungen mit Introcid bei fieberhaften puerperalen Enkrankungen. Medizinsche Klinik 25:587-588, 1928.

34.   Fox, C.L.: Silver sulfadiazine. A new topical therapy for pseudomonas in burns. Archives of Surgery 96:184-188, 1968.

35.   Monafo, W.W.: Personal communication

36.   Monafo, W.W., Tandon, S.N., Ayvazian, V.H.,Tuchschmidt, J., Skinner, A.M. and Deitz, F.: A new topical antiseptic for extensive burns. Surgery 80:465-473, 1976.

37.   Fox, C.L., Monafo, W.W., Ayvazian, V.H., Skinner, A.M., Modak, S., Stanford, J. and Condict, C.: Topical chemotherapy for burns using cerium salts and silver sulfadiazine. Surgery, Gynecology & Obstetrics 144:668-672, 1977.

38.   Saffer, I.D., Rodeheaver, G.T., Hiebert, J.M. and Edlich, R.F.: In vivo and in vitro antimicrobial activity of silver sulfadiazine and cerium nitrate. Surgery, Gynecology & Obstetrics 151:232-236, 1980.

39.   Munster, A.M., Helvig, E. and Rowland, S.: Cerium nitrate-silver sulfadiazine cream in the treatment of burns: a prospective evaluation. Surgery 88:658-660, 1980.

40.   Helvig, E.I., Munster, A.M., Su, C.T. and Oppel, W.: Cerium nitrate-silver sulfadiazine cream in the treatment of burns: a prospective randomized study. The American Surgeon 45:270-272, 1979.

41.   Bowser, B.H., Caldwell, F.T., Cone, J.B., Eisenach, K.D. and Thompson, C.H.: A prospective analysis of silver sulfadiazine with and without cerium nitrate as a topical agent in the treatment of severely burned children. The Journal of Trauma 21:558-563, 1981.

42.   Holder, I.A.: In vitro inactivation of silver sulphadiazine by the addition of cerium salts. Burns 8:274-277, 1982.

43.   Allgower, M., Cueni, L.B. and Stadtler, K.: Burn toxin in mouse skin. The Journal of Trauma 13:95-111, 1973.

44.   Peterson, V.M., Hansbrough, J.F., Wang, X.W., Zapata-Sirvent, R and Boswick, J.A.: Topical cerium nitrate prevents postburn immunosuppression. The Journal of Trauma 25:1039-1044, 1985.

45.   Allgower, M.: Prologue. Burns 21(suppl 1) S5-S6, 1995.

46.   Allgower, M. Schoenberger, G.A. and Sparkes, B.G.: Burning the largest immune organ. Burns
       21(suppl. 1):S7-S47, 1995.

47.   Burns 21(Suppl. 1) 1995.

48.   Dömötör, E.: Treatment of first and second degree burns with phlogosam ointment. Ther. Hung
       17:40-43, 1969.

49.   Guidi, G.: Contributo alla farmacologia delle terre rare: il neodimio. Arch. Int. Pharmacodyn.
       Ther. 37:305-348, 1930.

50.   Mancini, M.A.: Contributo alla farmacologia delle terre rare III- azione astringente dei sali di
       lantano. Archivio di Fisiologia 25:257-271, 1927.

51.   Dyckerhoff, H. and Goosens, N.: Über die thromboseverhütende. Wirkung des neodyms
       (Neodympräparat Auer 144") Zeitschrift für Exp. Med. 106:181-192, 1939.

52.   Vincke, E.: Die blutgerinnungshemmende wirkung der seltenen Erden. Hoppe-seyler's
       Zeitschrift für Physiologische Chemie 272:65-80, 1941.

53.   Vincke, E. and Schmidt E. Über den Prothrombimangel durch seltene Erden und seine
       Beeinflussung diurch Vitamin K₅. Hoppe-Seyler's Zeitschrift für Physiologische Chemie 273:39-
       46, 1942.

54.   Vincke E. and Oelkers, H.A. Zur pharmakologie der seltenen Erden. I. Mitteilung: Wirkung auf
       die Blutgerinnung. Archiv. Für Exper. Path. Und Pharmakol. 187:594-603, 1937.

55.   Beaser, S.B., Segel, A. and Vandam, L. The anticoagulant effects in rabbits and man of the
       intravenous injection of salts of the rare earths. Journal of Clinical Investigation 21:447-454,
       1942.

56.   Vincke, E. and Sucker, E.: Über ein neues antithromboticum das neodymsulfoisonicotinat.
       Klinische Wochenschrift 28:74-75, 1950.

57.   Thies, H.A. and Boecker, D.: Klinische Erfahrungen mit einen neuen Antithrombotikum ans der
       Gruppe der seltenen Erden. Deutsche Medizinische Wochenschrift 78:222-224, 1953.

58.   Wilbrand, U.: Klinische Erfahrungen mit den neuen Antikoagulans Thrombodym. Deutsche
       Medizinische Wochenschrift 78:330-332, 1953.

59.   Vincke, E.: Thrombosephrophylaxe mit Thrombodym. Therapie der Gegenwart 11:509-516,
       1960.

60.   Hunter, R.B. and Walker, W.: Anticoagulant action of neodymium-3-sulpho-isonicotinate.
       Nature 178:47, 1956.

61.   Divald, S. and Joullie, M.M.: Coagulants and Anticoagulants. in Medicinal Chemistry (Burger A.
       Ed.) pp. 1092-1122. 3rd edition, part II Wiley-Interscience,New York, 1970.

62.   Nath, B.B., Chattopadhyay, S. and Sarkar, S.: Lanthanides as blood anticoagulants: Mode of
       Action. Indian Journal of Biochemistry and Biophysics 19:32-36, 1982.

63.   Funakoshi, T., Furushima, K., Shimada, H. and Kojima, S.: Anticoagulant action of rare earth metals. Biochemistry International 28:113-119, 1992.

64.   Jancso, N.: Inflammation and inflammatory mechanisms. Journal of Pharmacology 13:577-594, 1962.

65.   Szporny, L. Cited by Domotor, ref. 67.

66.   Balogh, G.: Phlogosol therapy in inflammations of the oral mucosa. Ther. Hung. 22:83-89, 1974.

67.   Based upon information provided by G. Lazar, personal communication.

68.   Hisaden, K. and Ando, A.: Radiolanthanides as promising tumor scanning agents. Journal of Nuclear Medicine 14:615-617, 1973.

69.   Havron, A., Davis, M.A., Selter, S.E., Perkins-Hurlburt, A.J. and Hessel, S.J.: Heavy metal particulate contrast materials for computed tomography of the liver. Journal of Computer Assisted Tomography 4:642-648, 1980.

70.   Brady, T.J., Goldman, M.R., Pykett, I.L. Buonanno, F.S., Kistler, J.P., Newhouse, J.H., Burt, C.T., Hinshaw, W.S. and Pohost, G.M.: Proton nuclear magnetic resonance imaging of regionally ischemic canine hearts: effect of paramagnetic proton signal enhancement. Radiology 144:343-347, 1982.

71.   Caillé, J.M., Lemanceau, B. and Bonnemain, B.: Gadolinium as a contrast agent for NMR. American Journal of Nuclear Medicine 4:1041-1042, 1983.

72.   Brasch, R.C., Weinmann, H.J. and Wesbey, G.E.: Contrast-enhanced NMR imaging: animal studies using gadolinium-DTPA complex. American Journal of Roentgenology 142:625-630, 1984.

73.   Carr, D.H., Brown, J., Bydder, G.M., Weinmann, H.J., Speck, U., Thomas, D.J. and Young, I.R.: Intravenous chelated gadolinium as a contrast agent in NMR imaging of cerebral tumours. Lancet 1:484-486, 1984.

74.   Carr, D.H., Brown, J., Bydder, G.M., Steiner, R.E., Weinmann, H.J., Speck, U., Hall, A.S. and Young, I.R.: Gadolinium-DTPA as a contrast agent in MRI: Initial clinical experience in 20 patients. Americal Journal of Roentgenology 143:215-224, 1984.

75.   Wesbey, G.E., Higins, C.B., McNamara, M.T., Engelstad, B.L., Lipton, M.J., Sievers, R., Ehman, R.L., Lovin, J. and Brasch, R.C.: Effect of Gadolinium-DTPA on the magnetic relaxation times of normal and infarcted myocardium. Radiology 153:165-169, 1984.

76.   Wolf, G.L. and Fobben, E.S.: The tissue proton T_1 and T2 response to gadolinium DTPA injection in rabbits a potential renal contrast agent for NMR imaging. Investigative Radiology 19:324-328, 1984.

77.   Runge, V.M., Clanton, J.A., Lukehart, C.M., Partain, C.L. and James, A.E.: Paramagnetic agents for contrast-enhanced NMR imaging: A review. American Journal of Roentgenology 141:1209-1215, 1983.

78.   Runge, V.M., Clanton, J.A., Herzer, W.A., Gibbs, S.J., Price, A.C., Partain, C.L. and James, A.E.: Intravascular contrast agents suitable for magnetic resonance imaging. Radiology 153:171-176, 1984.

79.   Goldstein, E.J., Burnett, K.R., Hansell, J.R., Casaia, J., Dizon, J., Farrar, B., Gelblum, D. and Wolf, G.L.: Gadolinium DTPA (an NMR proton imaging contrast agent): chemical structure, paramagnetic properties and pharmacokinetics. Physiological Chemistry and Physics and Medical NMR 16:97-103, 1984.

80.   Gries, H., Rosenberg, D. and Weinmann, H.J.: Diagnostic Media. U.S. Patent 4,647,447. March 3, 1987.

Index

—C—

—D—

—I—

—H—

Chemists and Chemistry

1. M. Morselli: *Amedeo Avogadro* [1776–1856]. A Scientific Biography. 1984
   ISBN 90-277-1624-2
2. F.W.J. McCosh: *Boussingault* [1802–1887]. Chemist and Agriculturist. 1984
   ISBN 90-277-1682-X
3. D.A. Stansfield: *Thomas Beddoes, MD (1760–1808)*. Chemists, Physician Democrat. 1984   ISBN 90-277-1686-2
4. Pieter Eduard Verkade (1891-1979): *A History of the Nomenclature of Organic Chemistry*. Edited by F.C. Alderweireldt, H.J.T. Bos, L. Maat, P.J. Slootmaekers and B.M. Wepster. 1985   ISBN 90-277-1643-9
5. A. Thackray, J.L. Sturchio, P.T. Carroll and R. Bud: *Chemistry in America, 1876–1976*. Historical Indicators. 1985
   ISBN 90-277-1720-6; Pb 90-277-2662-0
6. G.B. Kauffman (ed.): *Frederick Soddy (1877–1956)*. Early Pioneer in Radiochemistry. 1986   ISBN 90-277-1926-8
7. R.B. Seymour and Tai Cheng (eds.): *History of Polyolefins*. The World's Most Widely Used Polymers. 1986   ISBN 90-277-2128-9
8. J.T. Stock and M.V. Orna, OSU (eds.): *The History and Preservation of Chemical Instrumentation*. 1986   ISBN 90-277-2269-2
9. N.A. Peppas (ed.): *One Hundred Years of Chemical Engineering*. From Lewis M. Norton (MIT, 1888) to Present. 1989   ISBN 0-7923-0145-8
10. R.B. Seymour (ed.): *Pioneers in Polymer Science*. 1989
    ISBN 0-7923-0300-8
11. H. Benninga: *A History of Lactic Acid Making*. A Chapter in the History of Biotechnology. 1990   ISBN 0-7923-0625-2
12. R. Mierzecki: *The Historical Development of Chemical Concepts*. 1991
    ISBN 0-7923-0915-4
13. S. Ross: *Nineteenth-Century Attitudes: Men of Science*. 1991
    ISBN 0-7923-1308-9
14. Torbern Bergman (1775) (ed.): *Chemical Lectures of H.T. Scheffer*. Translated by J.A. Schufle. 1992   ISBN 0-7923-1760-2
15. C.H. Evans (ed.): *Episodes From the History of the Rare Earth Elements*. 1996
    ISBN 0-7923-4101-5

KLUWER ACADEMIC PUBLISHERS – DORDRECHT / BOSTON / LONDON